Myths are ostensibly 'true', that is, they present themselves as giving an accurate narrative of 'what really happened'. A culture rarely recognises its own mythology as mythology. Judged from within a culture, myths are true accounts of the way things really are.

Professor Elizabeth Vandiver, *Classical Mythology*

OUR GODLESS UNIVERSE

Natural Selection and the Spiritual Search for Meaning Without Religion or Supernaturalism

TONY SUNDERLAND

OUR GODLESS UNIVERSE

Natural Selection and the Spiritual Search for Meaning Without Religion or Supernaturalism

ISBN: 9798720666040

Published by Sunderland Education for Amazon.

Author's Note

For the past 20 years, I have investigated the history and myths of Western civilisations in an attempt to make sense and create meaning in a world that, for many of us, appears senseless and meaningless. At the least, I gained some understanding of how past cultures endured the struggles life presented to them. But one thing gradually became starkly clear: when attempts are made to 'civilise' the way in which large groups of people live and die together, humans are inherently unstable and chaotic. Every ancient religion – and from this, every ruling human elite – was (and is) hell bent on one thing: ensuring order over chaos. What 'order' represents is determined solely by those individual(s) who hold sway over the rest of us.

Large groupings of people living in a confined environment is a relatively new phenomenon in human history. For hundreds of thousands of years, we (and our earlier hominid relatives) lived in small tribal groups that were mostly formed along lines of kinship and dominated by an alpha male. Life was short and frighteningly brutal. The sheer terror that darkness brought at the end of each day would only end with the dawn of the resurrected sun. Who we were, and the genes that have passed on to us from the ancient past, are still very much part of who we are, even today. Therefore, who we are now is heavily affected by who we were in our deep past. The genetic traits that helped us survive and reproduce in our very ancient past are still deeply embedded in every person alive today. Given the complexity of the world we live in now, these anciently sourced traits may not always act in the best interests of how we wish to interact with others, and indeed how we wish to interact with our own inner self.

My claim that we are a violent and unstable species – particularly when confined to large groupings – may, for some people, be both confronting and even false. But if we open any history book, written from any perspective, we find violence, deception, murder, and betrayal placed front and centre of each page. If we dare to be honest with ourselves, most of us would also admit to the suppression of selfish and even violent impulses that continually make their presence known deep inside us. These urges are derived from ancient genetic traits which instinctively fight for survival. Personally, however, it seems that most of us cannot bear very much reality and quickly deny our chaotic natures.

Paradoxically, it appears that the innate cause of our inherent instability, the societies and cultures we have created to house large groupings of people, may also be the pathway for escaping – or at least managing – our genetically predetermined nature. In other words, our tendency for violence and aggression may be suppressed by the civilising effect of our cultural institutions. It must be acknowledged, however, that this trait for violence and aggression is universal, and therefore beyond any artificially created notions of race or nationality. In fact, if we again open the books detailing our history, it is immediately clear that any institution, race or national cause claiming some sort of authoritative high moral ground is most likely to be among the most tyrannical, treacherous, and hypocritical culprit of its time.

That said, we are also the only species which has evolved to the point where the confines of natural selection are being breached to the extent that we are capable of intelligently designing ourselves and our future. It is important to stress that my use of the term 'intelligent design' does not include any

conceptualisation of a supernatural deity or divine designer. Rather, when used in its proper form and context, *intelligent design* is the deliberate intervention of intelligent beings in the natural order of things; because of this, certain features in the universe are best explained by an intelligent cause. I believe that evolution by natural selection and intelligent design are not mutually exclusive concepts; rather, they are both compatible and interdependent in explaining where we came from and why we are here.

This book is about the collective journey we have taken so far, albeit from an exclusively Western perspective. The road I travel will assume that there is no supernatural entity either guiding or misleading us; the fog of superstition does indeed obscure the contours of truth. The story told here is in two parts: First, I discuss the unbroken narrative of our genes across time, and the cultural attempts to transcend our predetermined genetic nature – both personal and societal. In the second part, I will discuss the constant spiritual yearning that resides within each of us, and that urges us to envision more than mere survival. But this impulse for spiritual yearning – unique among human beings – has been abused and manipulated by other (usually religiously inspired) forces. I will assume that, across time and space, there may be some method in the madness we have endured along the way; and it begins with the most powerful organising force in the universe: evolution by natural selection.

Figure 1: The Great Rift Valley as viewed between Israel and Jordan. This 6000 km trench begins in Mozambique and ends just above modern-day Lebanon. It is widely believed that the first humans to emerge from Africa made their way through this ancient landform. The upper part of the rift was also home to the first monotheistic religions. It is, therefore, a place where science, evolution, and religion have all made their collective mark on the story of humankind.

CONTENTS

PART A: A UNIQUE CREATION

In the distant future I see open fields for far more important researches. Psychology will be based on a new foundation, that of the necessary acquirement of each mental power and capacity by graduation. Light will be thrown on the origin of man and his history.

Charles Darwin, *On the Origin of Species*

ONE

All in the Genes

Evolution is a fact. Beyond reasonable doubt, beyond serious doubt, beyond sane, informed, intelligent doubt, beyond doubt evolution is a fact.

Richard Dawkins, *The Greatest Show on Earth*

Introduction

A few years ago, my partner gave me a birthday present: a genetic testing kit. Excited by the prospect of finding out something historically tangible about the man she had married 25 years ago, she duly stood by while I filled a vial with saliva, and then carefully packed it and posted it off. She was certain that my apparent inner turmoil existed because my Irish genes were in constant conflict with the English part of me! Within four weeks we received a detailed history of my genetic past and links to people to whom I am related. I was astonished to see the sheer diversity of cultural backgrounds that were presented in the results. I was also left with the disturbing (and also immensely interesting) fact that who I am is part of an ancient, genetically linked chain of life that has survived our chaotic past. Looking at the piece of paper in my hand, another unifying thought came to mind: that all of us are born uniquely burdened by our genes.

Most of us have heard of the terms evolution and natural selection, but few people take the time to consider deeply the significance of these processes in determining who we are, where we came from, and indeed how we live our lives. I have purposely structured this chapter to hopefully explain some of the more important ideas behind evolutionary theory. It is also important to state that evolution by natural selection does not reflect the intrinsic beauty we can experience through the natural world; in reality, natural selection is a ruthless and amoral mechanism of survival.

The first section will introduce the mechanics of biological natural selection, with emphasis on our own species. A brief discussion will also be included on some of the more controversial realities presented to us by the newfound ability of science to, in some degree, transcend the confines of Darwinian evolution. These new realities cannot be ignored or denied. We now have the capability to design new and more perfected (or preferred) versions of our species. It is not surprising that the chapter following this one will concentrate on the moral landscape of this insecure new world. It is also important to clearly state that I believe any decisions regarding genetic modifications of the human form should be ideally placed with in the domain of personal agency. State legislation has the potential for totalitarian rule, and should only be considered as a last resort in circumstances where individual and communal well-being are threatened. Religious interventions must be ignored and discarded as irrelevant, and possibly malicious.

More than a Theory

Who we are is the result of millions of years of evolution. Our ancient inherited genetic code has been finely tuned by natural selection, and it is a testament to the success of cumulative adaptions that we have survived and prospered as a species. For better or worse, we are without doubt the most dominant organism on Earth today. Homo sapiens – Latin for "wise human" – emerged around 200,000 years ago. In form and physical function, we are still in essence this same descendant from our ancient primate relatives. However, each of us is also a unique product of evolution. The subsequent diversity of genetic traits between individuals, and the influence of these personally inherited traits on how we experience the world around us, transcends all other influencing factors. This one salient fact is the overriding premise on which this work hinges. Who we are, where we come from, and where we are going is all part of one narrative – the journey of our genes through what can be termed *deep time*.

The theory of evolution through natural selection was pioneered by Charles Darwin and Alfred Russel Wallace. But it was Darwin's 1859 book *On the Origin of Species* that inspired a revolution in how we thought about our past and how we came into being. Darwin demystified the greatest mystery – where do we come from? We now know, through scientific evidence, that the earth is around 4.6 billion years old and that the first living organism – our common ancestor – appeared around 3.8 billion years ago. If taken to its end point, Darwin's theory of evolution suggests that all known species are descended from this common ancestor. Over this extremely long period of time, basic forms of life, such as bacteria, evolved into the grand mosaic of life we see on Earth today. **Darwin's original theory** was later refined to include the work of Gregor Mendel and the specific role of genes in natural selection, sometimes known as Neo-Darwinism.

Biological evolution refers to the change in inherited traits of a population from generation to generation. Specific traits are found in the expression of genes (the units of evolution) that are copied and passed on to offspring during reproduction. Evolution can occur either randomly, through mutations and genetic drift (changes in gene frequency within a population), or in non-random ways through natural selection. Evolution through natural selection is an amoral, mechanistic process of survival of those organisms best suited to the environment in which they live. Only those organisms that survive long enough to reproduce can pass on their genetic traits to future generations.

Natural selection mostly occurs in slow increments over exceedingly long periods of time, and is dependent on reproduction processes that mix the genes of two organisms, almost exclusively through sexual reproduction within a species. In terms of the long-range ability of a species to survive the rigours of evolutionary natural selection, the continual changing circumstances of sexual reproduction are superior to asexual forms of reproduction. Sexual reproduction – where two different organisms create new offspring – appeared approximately 2 billion years ago.[1] Reproduction in this way results in a unique creation, a zygote, which is a combination of the genetic material of each parent. In contrast, asexual reproduction results only in an exact replication of the original organism, and subsequently offers no chance of adaptation through the rigours of natural selection. Interestingly, certain types of bacteria (prokaryotes) obtain genetic variation through a non-sexual unilateral exchange of genes.

Although only half the age of the earth, sexual reproduction is the ancient, driving force that has created increasingly complex organisms. **The cellular chaos of sexual reproduction can also create positive mutations that can survive and adapt to changing environments. Sexual reproduction dilutes 50 percent of each parent's genetic contribution to their offspring.** It is important to note that the newly inherited mix of genes does not mean that the genes are blended together, but rather are equally distributed intact – half from each parent. At birth, our genes are not a 'melted' outcome, but rather a 'reshuffled' result of the genes of each parent. Those offspring with gene variants that best suit their environment and reproductive chances are 'naturally' selected to survive. Advantageous inheritable traits are then passed to the next generation.

The primary focus of evolution is therefore the passing on of genes. There is no God-inspired plan – only the relentless fight for survival through natural selection and reproduction. Genes merely need to be perfect enough to survive and reproduce. Unfortunately, this amoral and automatic struggle for survival in natural selection can also be unkind. Sexual selection, a special form of natural selection, indicates that the chances of a sickly, short, balding, overweight male passing on his genes are less than those of a stronger, taller, alpha leader, given that other determinants, such as income and social status, are equal. The laws of natural selection make this choice automatic. It is estimated that over eight percent of males spanning sixteen different populations in Asia are direct descendants of the Mongol leader, Genghis Khan – rumoured to be tall, dark and very handsome!

Interestingly, one of the more puzzling and controversial dilemmas with regard to the process of natural selection through sexual reproduction is the persistence of male homosexuality throughout humanity and in other parts of the animal kingdom. At first glance, it would be reasonable to conclude that the genes of homosexuals are less likely to be passed on, and that because of this, any inherited trait for homosexuality should diminish over time. This has not happened. But in recent years, new research offers a variety of theories that point to the conclusion that a genetic disposition for homosexuality can also be embedded in our genes, through an indirect journey along the X chromosomes of maternal relatives.

One of the more convincing arguments for this premise comes from Professor Andrea Camperio–Ciani, located at the University of Padova in Italy. Her research indicates that the "maternal female relatives of gay men have more children than maternal female relatives of straight men. The implication is that there is an unknown mechanism in the X chromosome of men's genetic code which helps women in the family have more babies but can also lead to homosexuality in men." [2]

Every person who enters this world is a unique representation of the fruits of natural selection. Our communal survival always has been, and most probably always will be, dependent on our species being extremely diverse and adaptive to changing circumstances. It is unlikely that the author of this book would survive the harsh reality of Stone Age life – but my ancient ancestors did! This is probably one of the reasons that people are now interested in finding out details of their genetic past and participating in DNA ancestry testing. Many of us have also abandoned notions of a heavenly father and instead want to connect with our actual great-great-great-grandmother.

The evolution of some genetic traits can also result in the formation of biological spandrels. A spandrel can be defined as a by-product of adaptive phenotypic (observable) traits. These traits are relatively benign in that they offer no adaptive advantage, but continue to exist because they present no harm to the organism. Probably the most obvious spandrel is the continued presence of nipples on human males. That said, spandrels may have an indirect environmental influence. For example, some traits considered 'useless' from an evolutionary perspective, such as musical skill, may influence non-evolutionary aspects within a cultural environment. Our enduring capacity for self-deception, while it offers no specific adaptive advantages, may be another example of a spandrel.

Figure 2: Charles Darwin (1809-1882) revolutionised how we view ourselves and our position in the world around us. For Darwin, it is the sheer diversity of life, and the constant struggle to survive and reproduce, that is at the heart of evolution through natural selection.

Selection Vehicles and Replicators

We know that all matter is composed of atoms. There are a finite number of atoms that in turn make up the 118 known elements – the building blocks of physical matter. Elements can combine to form molecules. Living organisms are mostly composed of macro (large) molecules. DNA (deoxyribonucleic acid) is a macro molecule that provides the digital code by which certain proteins influence cells in developing embryos. Each 'chain' of DNA contains two copies – one from each parent – of 23 chromosomes. Genes are those parts of DNA that define the traits an organism will inherit from their parents. It was only during the middle of the 20th century that our 2-billion-year-old genetic code was deciphered by James Watson and Francis Crick. Each person's unique genome is a quaternary (as opposed to binary) string of data. Each code particle (adenine, thymine, cytosine, and guanine) – shortened to the letters A, T, C, and G – can be written down as individual units of data. *A* can only pair with *T*, and *G* can pair only with *C*. These are known as our genetic 'base pairs.' The nucleus of nearly every cell in the human body contains our entire genetic blueprint in its DNA.

DNA has a dual purpose: to act as a template for replication and a biological blueprint for the construction of life. The human genome consists of some three billion nucleotide base pairs. Within this, there may be up to around 21,000 protein-coding genes.[3] However, at a cellular and molecular level, there are design flaws within many bio-chemical pathways. The reassortment of chromosomes during sexual reproduction only further increases the chances of functional failure even outside the pressures of natural selection, with the result being the death of many newly formed zygotes and embryos. This chaotic process is far from any religiously inspired, divine design by an all-loving supernatural creator. Only recently have scientists realised the extent of inherent instability in replicating human genes. For example, around five percent of humanity have obvious genetic disorders and around 20 percent of infant deaths are from genetic causes. Approximately 50 percent of all miscarriages are due to genetic flaws.[4]

Everything we are, and where we come from, is embedded in our genetic code. After our death, two things survive us: our genes and our memes. Richard Dawkins first uses the term *meme* in his groundbreaking book *The Selfish Gene,* in 1976. He uses this word to describe enduring contributions to culture, and the lived examples of life that survive at least one generation into the future. It should be noted that Dawkins' idea of a cultural meme is vastly different than 'internet memes,' which are usually humorous and transient. For Dawkins, memes are significant cultural messages that can impact the way a society evolves over time. That said, the only thing truly immortal about any human is his or her genes. They are the only things that can survive us through our offspring. There are 'two units of selection:' vehicles (the living organisms that survive long enough to pass on genetic information) and replicators (the actual genes that have been delivered by the vehicle).

It is also important that we are aware of and accept the fact that we are the product of thousands of years of chaotic and violent battles of survival. Natural selection by definition discriminates against those organisms that have not adapted successfully to an environment. Dawkins, in his book *Science in the Soul*, succinctly summarises this premise as Darwinism – "the non-random selection of randomly varying replication entities by reason of their phenotypic effects" – as the only valid explanation for the evolution of adaptive, complex organisms.[5] Every evolved individual is unique, but at the same time each is linked to the greater biopopulation of humanity.

Most people also understand Darwinism and natural selection as being "the survival of the fittest." This term was originally used by Herbert Spencer in 1852 after he had finished reading Darwin's book. Darwin subsequently also used the term as his preferred way of explaining the amoral workings of natural selection, where only those organisms best adapted to their immediate environment will be most likely to survive and reproduce. Darwin himself conceded that "the survival of the fittest" better expressed the non-random and cutthroat environment which enables evolution over time.

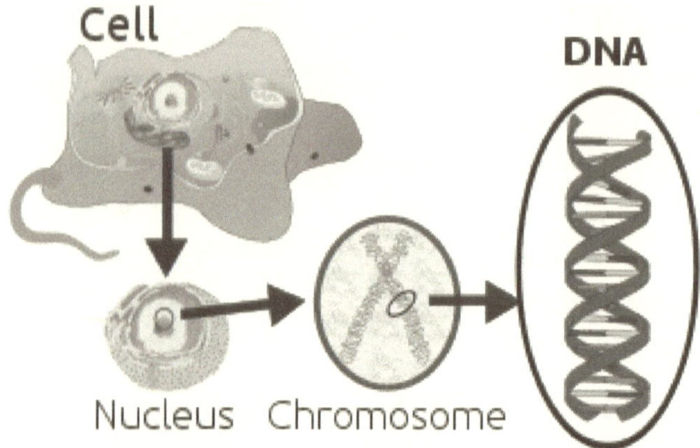

Figure 3: Working backwards, our DNA, as it appears contained within a double helix, exists on either the X or Y chromosome, which in turn resides in the nucleus of a living cell - the smallest fundamental unit of structure and function in living organisms.

An Uncomfortable Truth

There is an increasing amount of evidence inferring that our genes also dictate who we are and how we live our lives. In other words, genetic determination may be the primary catalyst for not only the human form but also for human endeavour and behaviour. Genetic determination controversially claims the high ground where nature (our genetic makeup) dominates nurture (our cultural environment). Even our cultural contexts are created by groups of 'genetically formed individuals.' Therefore, any environmental influences on a person's behaviour are also causally linked to the genetically determined traits that natural selection has influenced over time. This fact severely blurs any nature/nurture dichotomy. As author John

Avise puts it when referencing groundbreaking studies of identical (monozygotic) and fraternal twins:

> For almost every behavioural trait so far investigated, from reaction time to religiosity, an important fraction of the variation among people turns out to be associated with genetic variation. This fact needs no longer be subject to debate; rather, it is time instead to consider its implications.[6]

The nature vs. nurture argument was more broadly investigated in a 30-year study of 1,000 people residing in Dunedin, New Zealand. This research is significant because it managed to retain over 90 percent of research participants from infancy to adulthood. Its conclusions are controversial because it showed, beyond reasonable doubt, that how we turn out during the journey of our life is significantly predetermined by our genes. One of the authors of the study, Terrie Moffitt, says, "All people are not created equal. Some have real gifts and talents, and some have real problems right out of the starting block. Once we accept that, we can't dodge the responsibility for social action."[7] Interestingly, the study also concluded that although a person may have a specific genetic disposition – say, for brutality or criminal behaviour – it is not always acted out on. A person's upbringing (nurture) in some instances suppressed the subject's genetic disposition. Nurture is not just the intimate bonds and relationships formed between people; it is also part of the greater societal context that surrounds a person. Our past no longer totally dominates what we become. Nature (our genes) may load the gun of our behaviour, but nurture gives us the ability to not pull the trigger, or to aim our behaviour in alternative directions.

Environmental factors also include the actions of chemical compounds and proteins (known as our epigenome) that can attach to DNA and direct processes which can turn genes on or off. These epigenomic compounds attach to DNA and modify its function. For example, women who are given folic acid supplements during pregnancy have a reduced risk of their foetus developing spina bifida. What we drink, eat, or even smoke may also 'pull the trigger' of a particular gene(s). For example, it is now generally accepted that some individuals who possess higher amounts of COMT genes are more prone to hallucinations and schizophrenia after smoking marijuana.[8] That said, it must be conceded that our genes really do load each of us with predetermined, and finite, abilities and flaws. Given the brave new world of genetic sequencing and engineering, this emerging and uncomfortable truth should not be ignored.

Many of the conclusions found in the study of evolutionary biology and genetic science are indeed both startling and uncomfortable. But these truths explain 'what is' rather than 'what ought' to be. Most controversial is the science of eugenics – deliberate actions undertaken to change the genetic composition of an individual or population through selective breeding. Many actions taken by what could be termed 'negative eugenics' are widely sanctioned in the community, for example, the discarding of embryos indicating conditions such as Huntington's Disease in favour of healthier alternative embryos. However, the possible selection of only those embryos that promised higher intelligence, musical ability, and so on (positive eugenics) would almost certainly be rejected by many people.

Genetic testing of human foetuses gives potential parents an indication of the likelihood their child will be inflicted with a disease, either at birth or in later life. The selective termination of genetically disabled foetuses is now a central issue in making reproductive decisions. Whether we like it or not, these interventions are a form of positive eugenics. However, any 'new eugenics' must not resemble the horrors inflicted by the Nazi programs of the mid-20th century. John Avise convincingly argues that it should also not invoke any form of 'God-inspired' moral authority:

> The only wrong approach is that in which the moral authority of a God (or tyrant) is asserted. As judged by the diversity of opinions held by responsible individuals on ethical matters pertaining to the human condition, any supernatural deity either has been strangely silent on such issues or else has conveyed vastly different messages to different listeners.[9]

What should be universally recognised, however, is that positive eugenics is not only possible, but also a reality. Different cultures over long periods of time have encouraged some form of selective breeding, particularly with regards to domestic animals. This is obviously apparent in the variation we see in different breeds of dogs. Varying physical attributes, such as colour, size, and facial features, have been selectively bred in different breeds to the extent that, for some people, it is hard to believe that all dogs are descended from the wild wolf.

Certain advantageous traits, such as intelligence and personality dispositions, have also been intelligently – artificially and outside the confines of natural selection – designed. A passive and comforting King Charles Cavalier is vastly different to an aggressive but intelligent German Shepherd. This reality can be seen, touched, and heard in many Western households. The same thing is undoubtedly possible in humans. As Richard Dawkins concludes with respect to the possibility of selectively breeding for intelligence in humans: "I'm sorry to have to tell you that positive eugenics is not ruled out by science… It doesn't matter if we can't agree how to measure intelligence, we can breed for any of the disputed measures or a combination of them."[10]

What has changed is the ability of science to not only identify traits of specific genes; it now increasingly has the power to positively select and, in some circumstances, design desired traits. The question of how these discoveries should be used is presently outside the scientific jurisdiction of 'what is' and is firmly placed in the more abstract realm of 'what ought' to be. In recent years, a fringe group of molecular geneticists have been actively searching for outlier groups of people who have gene variants that display advantageous traits such as superior metabolism and resistance to type 2 diabetes. By analysing the way DNA works on these genes, they hope to design new medications and activation mechanisms that are the first steps in creating 'super humans' among the living. In 2018, Chinese researchers used a gene-editing tool called CRISPR– cas9 (clustered regularly interspaced short palindromic repeats) to remove a gene related to the HIV virus from at-risk embryos. Gene editing is controversial and condemned by many bioethicists because edits performed on sperm, eggs, or embryos can be passed on to future generations.[11]

People can now have their entire genome sequenced for less than a thousand dollars. Newborn babies are routinely screened (with parental permission) for a variety of genetically sourced illnesses. Parents, if they wish, can also obtain a detailed summary of what genetic gifts and challenges their child will face throughout life. This is not science fiction; it is a scientific fact. The United Kingdom has in place a plan to fully sequence all newborns as part of a universal cradle-to-grave healthcare system.[12] The moral challenges of this plan are significant. A baby cannot give consent and it appears that the wishes of parents are sidelined and subordinated by government control. A totalitarian regime in the future will have everything it needs to 'select' who is fit to live.

Gene sequencing of embryos also brings with it a new set of moral dilemmas. Whole genome sequencing (WGS) would allow parents to select an embryo that best suits their desires and aspirations for their child. For example, if they wish to produce offspring with extreme superior intelligence, they will only select those embryos that contain genetic variants in locus *ADAM12 gene* that are linked to enhanced cognitive abilities and educational attainment. If they wish to intelligently design their baby, then the newly discovered gene editing tool CRISPR may also be used.[13]

Up until the beginning of this century, humanity was a passive participant in the process of genetic evolution. However, innovations in molecular genetics have opened a new frontier where human engineering of genes is now a reality. It is now possible to genetically manipulate the human condition. Human cloning, genetic screening of disease, gene therapy, and sex selection are tangible realities. For the first time in history we can intelligently design the future of our species and be our own creator. It is therefore important to stress, yet again, that the science of evolutionary genetics is descriptive (what we do) as opposed to prescriptive (what we should do). Empirical science tells us what is, it does not inform us what we ought to do from any form of moral perspective.

It could also be argued that Darwinian natural selection – through survival and adaptation mechanisms – in humans has been somewhat neutralised. Most people born today will survive to an age when reproduction is possible; the majority of these people will be lucky enough to transfer their genes to the next generation. That said, all of us are still prisoners of our ancient, genetically determined dispositions. Given this new frontier of human empowerment, some critical questions arise. What aspects of our lives and our communities do we consider as worthwhile endeavours? How do we traverse this untrodden domain? And most importantly, what do we value?

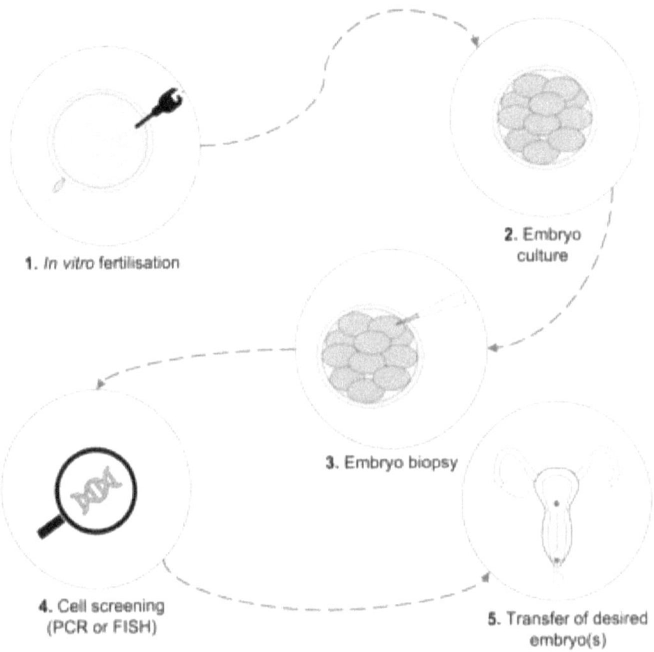

Figure 4: Pre-implantation genetic diagnosis (PIGD) is a form of genetic screening which involves the testing of human embryos prior to being transferred to the womb. This is especially useful when a parent has a heritable disease. PIGD can also be used for sex-selection purposes.

Key Points

- The evidence of evolution through natural selection is overwhelming, and who we are is the result of millions of years of evolutionary adaptations. Without exception, the primary focus of evolution is the passing on of genes. There is no God-inspired plan – only the relentless fight for survival through natural selection and reproduction.

- Our own species is the product of thousands of years of chaotic and violent battles of survival. Natural selection by definition discriminates against those organisms that have not adapted successfully to an environment.

- After our death, two things survive us: our genes and our memes. There is an increasing amount of evidence inferring that our genes play a major role in determining who we are and how we live our lives.

- Genetic determination may be a primary catalyst for not only the human form but also for human endeavour and behaviour. Nature loads the gun of our behaviour, but nurture has the ability to not pull the trigger.

- Many of the conclusions found in the study of evolutionary biology and genetic science are indeed both startling and uncomfortable. Most controversial is the science of eugenics. It must be recognised that it is now possible to genetically manipulate the human condition.

- Innovations in molecular genetics have opened a brave new world where human engineering of genes is now a reality. But these truths explain 'what is' rather than 'what ought' to be.

- Darwinian natural selection in humans has been somewhat neutralised. Most people born today will survive to an age when reproduction is possible, and the majority of these people will be lucky enough to transfer their genes to the next generation.

- For the first time in history we are now able to directly interfere with the mechanics of natural selection. Science has enabled us to intelligently design new versions of offspring. We cannot hide or deny this new, uncomfortable truth.

TWO

A Question of Values

A life cannot be thought to be good which is harmful in its impact on others, or which results in a perversion of the possibilities and capabilities of the person living it. A choice of life has to be able to stand up to scrutiny, a fact that rules out irresponsible, destructive, wasteful lives as good.

A. C. Grayling, *The God Argument*

Introduction

Given the breadth and complexity of this area of discussion, I have deliberately taken a position placed formally in the disciplines of evolutionary psychology and evolutionary biology, as my primary focus in investigating what we value and what we deem as being moral. By doing this, I have taken a compatibilist approach with regards to free will and determinism. Most of our base values are instinctually driven, but we have the ability to freely overcome or at least suppress impulses that may not benefit our societal existence. I concede that proposing a framework in which our values and desires are inherited through our genes is controversial, but a surprising amount of evidence points in this direction. Most of our base values are heavily skewed to our ancient past, and are therefore programmed primarily for survival and reproduction. Even acts of selfless sacrifice and charity may have their origins in the relentless but ordered chaos of natural selection. It is therefore important to distinguish between what is moral as opposed to what we value. An individual who knows the difference between right and wrong (from a societal perspective), and chooses right, is moral. A person's morality is therefore reflected in their willingness to do the right thing for their tribe/community. Values, in contrast, are personally conceived and acted on, but can also reflect a particular communal moral stance. Ethics are encoded moral values in action.

If many of our so-called moral decisions are made instinctively, it raises broader questions relating to human agency, free will and the role of our societal institutions in creating a moral society with established ethics designed for the good of all. I will attempt to provide both an explanation of the unstable reality of human nature, and outline some of the culturally-inspired ways in which our species is creating moral frameworks that reflect our emerging desire to transcend the confines of natural selection.

Core Values

A basic definition of the much-used term 'values' could simply be the principles an individual uses when deciding what is personally preferred, as opposed to what is rejected. Indeed, any organism that strives for something over the possibility of alternative options can be said to be 'value driven.' It could also be argued that a person's core values are embedded in their genes. Reward mechanisms (both positive and negative) are innate and moulded over time like other traits through the rigours of natural selection. For example, the values an individual (and their community) place on learning – and even what to learn – are ancient, primary traits passed on to ensure survival. The desire to learn is genetically encoded. Even the desire to communicate through language and the values we place on effective communication originate in the deep genetic structure of the human brain.[1]

Richard Dawkins further contends that what we assume is our discerning taste in art, music and food may also be genetically inherited. These genetically-constructed values and desires come from our ancient past. According to Dawkins, an environment of evolutionary adaptiveness and the conditions that surrounded it moulded the first values, and these ancient desires are still deeply embedded in our genetic makeup. As Dawkins says:

> Do not ask how a middle manager's ambitions for a bigger desk and a softer office carpet benefit his selfish genes. Ask, instead, how these urban partialities might stem from a mental module which was selected to do something else, in a different place and time. For office carpet, perhaps (and I mean *perhaps*) read soft and warm animal skins whose possession betokened hunting success.[2]

Research also indicates that the main categories of human disposition – aggression, religion, sex, and altruism – are also the result of evolutionary natural selection. Of these, the pursuit of altruism is obviously the hardest to reconcile against the selfish genes that continually fight for pre-eminence and immortality.[3] Altruism or, more basically, the urge to self-sacrifice can be divided into two forms of motivation: soft-core and hard-core altruism. The urge to self-sacrifice, from a soft-core perspective, includes charitable actions by an individual (or group) which will ultimately be rewarded by some form of reciprocation – you scratch my back, and I will scratch yours. This form of altruism is ultimately selfish and may have even influenced other genetically inherited traits for pretence, deceit, and hypocrisy. In contrast, hard-core altruism encompasses those actions of heroic self-sacrifice that may even end in death. This form of seemingly selfless altruism may be influenced by a desire to preserve the greater gene pool of kin and tribe. Acts of hard-core altruism confer 'fitness benefits' to close relatives and the tribal alliances that allow the genetic family grouping to survive.[4]

If we assume that the earliest humans lived in small tribal units, it is reasonable to conclude that these primitive environments of evolutionary adaptiveness encouraged the survival of genes that valued group and kin survival. Our selfish genes not only value us, but they also place value on kin and tribal survival. Kin selection theory expresses the idea that the selfish desire for personal gene perpetuation is extended to tribal kin. In smaller, tribal communities, people were not only dependant on each other; they were also genetically related. The value of altruism probably originated here. **As Darwin famously concluded:**

> there can be no doubt that a tribe including many members, who, from possessing in a high degree the spirit of patriotism, fidelity, obedience, courage, and sympathy, were always ready to give aid to each other and to sacrifice themselves for the common good,

> would be victorious over most other tribes;
> and this would be natural selection.[5]

Human emotions may have also evolved to serve the individual and the tribe in response to specific situations. Emotions such as fear not only give rise to physiological responses, but also automatically communicate to the group that danger is present.

It could be further contended that how much we value and trust another person is dependent upon the 'shadow of the future.' In other words, how much contact do we envision having with that person over time? There also appears to be a genetic tendency to discriminate against those people outside of our tolerance for 'the other.' It is starkly apparent that our genes have not had time to adjust to modern urban living conditions where tribal and even kin preference is blurred or even negated. That said, the persistence of hard-core altruism in most human cultures is a testament to the possibility that genes and cultures co-evolve. The cultural memes passed down to each generation show that although human nature may be genetically manufactured, it is also culturally inspired.[6]

Interestingly, Dawkins believes that the complexity of our evolved brains has the power to rebel against some aspects of our genetic nature. It is possible to construct our own (and communal) values that transcend our genetic drive for selfish survival. Social welfare and other community systems are testaments to the rebellion against our selfish genes. This act of rebellion is unique, but it must be stressed that even this ability is sourced in evolution and not divine law or intervention. That said, it is important that we are aware of and accept that we are the product of thousands of years of chaotic and violent battle for survival. Natural selection, by definition, discriminates against those organisms that have not adapted successfully to an environment. Unfortunately, our current genetically inherited values do not match with the sophisticated (and hopefully tolerant) cultures we have created. [7]

Optimistically, Dawkins hopes for a continuation in the unstable but positive "moral arc" towards a more compassionate and tolerant society. Although this arc is erratic and zigzagging, "the trend is unmistakably in one direction."[8] The cause for this is not adherence to any form of religious doctrine or dogma; rather, it comes from our own culturally-constructed desire for a better world – a meme communicated through our institutions, thoughts, and actions. At the heart of this progressive movement to a more moral society has been the acceptance of science and the empirical truths deducted by scientific research. There is an objective truth about the way things came to be (and are) in the world that surrounds us. Neo-Darwinism and the discovery of human origins give us an explanation of our deep past without resorting to supernatural explanations.

Figure 5: An Algerian cave painting dating from the Neolithic period, depicting a hunting scene. Early tribal communities were usually aligned through kinship, with a dominant alpha male ensuring both kin survival and reproduction of his own selfish genes.

Creating a Moral Framework

The interplay between human behaviour, our genes, and cultural environments has influenced the emergence of a relatively new field of scientific endeavour: evolutionary psychology. Broadly defined, evolutionary psychology is the formal investigation of the human mind, from a perspective that puts our genetic makeup as the centrepiece of determining human behaviour, much in the same way that evolutionary biologists put our genes at the forefront in determining the human form. Any so-called 'values-based' emotions which impact human behaviour originated from naturally selected adaptations in our deep past.[9] We still leverage these ancient traits today when making decisions. For example, when confronted with something that offends or disgusts us, we emotionally react against the transgression. Only after this reaction do we rationalise and justify our actions. In other words, we feel disgusted, act on our feeling, and then rationalise (defend our actions) morally. Professor Diana Fleischman has conducted extensive research on the adaptation of disgust as a human trait in response to the threat of disease. It is no coincidence that when someone vomits in our presence, the urge to respond in like fashion is almost overwhelming.[10]

The survival trait of disgust has now evolved to respond to situations where more abstract 'moral rules' are transgressed. For example, the emotional trigger for disgust is activated, for many people, when their national flag is disrespected or burned. This trait is also present when we feel other anciently-sourced emotions such as guilt and shame: we tend to first react emotionally, and only then rationalise the response that follows. Subsequent violent or submissive responses may exceed societal/cultural sensibilities, and therefore be deemed inappropriate. Because the trait of disgust is embedded in ancient survival instincts, there is a tendency for our emotions to 'misfire' and respond irrationally to some situations. This maladaptation poses a further dilemma. How can we improve our communal desire to act morally and overcome our evolved, animal impulses? Maybe the first step is to recognise that they exist within all of us. As Fleischman puts it, "One prominent criticism of evolutionary psychology is that it condones immoral behaviour by pointing to its natural origins.

Far from it, I believe we must acknowledge the shadow of our evolved morality in order to transcend it."[11]

The idea that our sense of morality can be sourced in a person's genetic past is controversial. It presupposes again that our nature is the prime driver of how we act and what we value. Traditional psychological constructs of morality are built on notions of human agency, where culturally-inspired nurture frames our individual and collective moral compass. Critics also argue that many of the hypotheses formed by evolutionary psychology are not falsifiable (that is, they can be proven right or wrong empirically) and therefore cannot represent valid scientific truth. Fleischman strongly disagrees with this premise and convincingly argues that evolutionary psychology is not only falsifiable, but should be the foundation for measuring improvements in our journey in becoming a moral species:

> We have made leaps and bounds in moral achievement compared to the deep history of humanity and to the rest of the natural world... The world is a much better place than it was, but an evolutionary perspective does predict limits to this moral expansion. The flourishing of sentient beings is still impeded by moral disgust, outrage, and self-deception.[12]

If we accept that our desire to be moral is genetically sourced from our ancient past, a new question arises: why is the trait enduringly beneficial? Evolutionary psychologist Oliver Scott Curry believes "that morality consists of a collection of biological and cultural solutions to the problems of cooperation recurrent in human social life."[13] It is this varied collection of cooperative traits that constitute morality. Across cultures and time these traits vary, but there appears to be a foundation of seven basic moral rules that are universal. He concludes:

> For 50 million years humans and their ancestors have lived in social groups. During this time natural selection equipped them with a range of adaptations for realizing the enormous benefits of cooperation that social life affords. More recently, humans have built on these benevolent biological foundations with cultural innovations – norms, rules, institutions – that further bolster cooperation. … And, as predicted by the theory, these seven moral rules – love your family, help your group, return favours, be brave, defer to authority, be fair, and respect others' property – appear to be universal across cultures.[14]

Layered over this conception of evolving human morality, it can be further contended that evolution has also installed within us a deep desire to be hyper-vigilant about harmful intent. Humanity has survived because it has adapted to form interdependent communities, where people need to cooperate openly and honestly. Each of us has an instinctual ability to read other individuals' mannerisms and actions for clues of possible intentions to harm the values and structures of their fellow tribe/community members. Given this, it is logical to conclude that something is moral if it reduces or minimises harm and promotes what is good. Our challenge now is, how do we find out if something or someone is 'good'?

Sam Harris, in his book *The Moral Landscape,* takes up this challenge. He first argues that something is moral and 'good' if it supports well-being.[15] Although well-being may be hard to precisely define, we can estimate if an action will either enhance well-being or diminish it. Harris places his trust in the objective truths found in science to give us a measuring gauge in creating a moral framework. Science, through its quest for truth and understanding, can provide facts of not only what is, but may also objectively guide us on the moral journey of "what we should do, and should want."[16] A critical point of delineation, however, is the premise that science does not create moral truths; it can only measure whether they are indeed truthful.

Harris then poses a critical question: How do we create and evaluate objective facts that can help discover the moral 'good'? As a starting point in finding objective moral truths, Harris first contends "that it is meaningful to ask whether a specific action or way of thinking will affect a person's well-being and/or the well-being of others."[17] The 2020 Covid pandemic was a rare occasion where many countries committed to personal self-isolation. Although most of the self-isolation programs were compulsory, those individuals who also voluntarily committed themselves to isolation and proper social distancing contributed significantly to reducing the spread of the virus. This new moral reality was initiated through scientific evidence. Interestingly, most denominations of religious faith also adhered to the restrictions placed on services and communal prayer gatherings. In the plague-ridden Middle Ages, people prayed to be spared; in 2020 they mostly adhered to the objective truth of science. The 'good' produced from self–

isolation, and adherence to scientific truth, most definitely enhanced collective well-being.

Harris also concludes that we must build better social structures and institutions "that reflects and enforces our deeper understanding of human well-being."[18] Science may aid in this search by at least attempting to objectively measure individual and collective well-being (the good). Central to this hope is the somewhat utopian ideal that individuals have the capacity to transcend their emotions and commit to objective reason. There may be such a time in the future, a time of "a science of morality."[19] Here, objective moral truths are hypothesised, tested, and discussed independently of unsubstantiated dogma or rules. These 'truths' will stand or fall based on their evidence. Unfortunately, it must be conceded that each person is a unique product of natural selection; we do not stand on equal ground. This reality cannot be denied. As Harris argues:

> No human being stands as author to his own genes or his upbringing, and yet we have every reason to believe that these factors determine his character throughout life. Our system of justice should reflect our understanding that each of us could have been dealt a very different hand in life. In fact, it seems immoral not to recognise just how much luck is involved in morality itself.[20]

As individuals, we are still instinctively driven towards the replication of our genes. Fortunately, our larger brains, social memes, and cultural traditions can rebel against the prison of our predetermined nature. We possess the ability to reject our destructive impulses; this is reflected in stable societies which reside outside the domain of natural selection. Socrates, the father of Greek philosophy, gave his life in sacrifice to the needs of his city-state ahead of his own selfish needs. This type of altruism is an enduring meme in Western cultures. It is also the beginning of an emerging cultural trait outside the laws of natural selection. As previously discussed, we alone are capable of rebelling against our genes. Richard Dawkins concludes that "We rebel when we give lectures, write books or compose sonatas instead of single-mindedly devoting our time and energy to disseminating our genes."[21]

Dawkins expands on this idea to speculate that an emerging trait of "super niceness" may be pivotal in directing humanity's moral arc towards "the good." Our emerging trait for "super niceness" may represent "a misfiring even a perversion of the Darwinian take on niceness."[22] This relatively new value originates from evolution of the human brain; it is not a divine gift or revelation from heaven. Interestingly, innovations in medicine, particularly over the past 100 years, have dramatically reduced infant mortality. Babies who would have perished in childbirth now survive. One of the main causes of death during childbirth has been the relative size of the baby's head in comparison to the mother's pelvis. Modern medical intervention has now largely solved this problem. This form of interventive selection may result in even bigger brains.

If science is to take a more active role in the search for a moral framework that recognises both our selfish nature and our enduring desire for something more than this, a new paradigm of belief may be needed. In the past, science has avoided taking any stance on the discovery and reality of moral truths. The moral relativism of submissive science has allowed faith-based religions to continue preaching mythologies about 'the soul' and the moral primacy of their exclusive beliefs. In contrast, science could join the rationalist quest to search for universally true answers to moral questions; we just need to discover them. To achieve this, we may have to revisit the relevance of out-dated philosophies and belief systems that have either been ignorant of, or have ignored or rejected, the uncomfortable truth that our genes are the driving force behind human behaviour and the human form.

Figure 6: Richard Dawkins is a highly recognised atheist, author and evolutionary biologist. His uncompromising stance on the dangers of organised religion has inspired many people to abandon their faith in supernaturalism and take personal responsibility for themselves and the world around them.

An Aggressive and Free Species?

As previously discussed, culturally-created beliefs and practices – and the memes communicated by them across time – are independent of the genes passed on by natural selection. We can now make choices that are from a perspective and position that are not solely determined by our genes. That said, it is apparent that the co-evolutionary soup of genes and culture has produced a variety of physical and cultural worlds that have uniquely evolved to suit varied selection requirements. But one thing remains constant in all societal creations: we are an innately unstable and potentially violent species. Sigmund Freud, in his pioneering work *Civilisation and its Discontents,* concluded that "the greatest hindrance to civilisation ...is the inclination of human beings to be aggressive towards one another."[23]

Freud also predicted there would be a crisis of identity in the Western world. He states: "the two processes of individual and of cultural development must stand in hostile opposition to each other and mutually dispute the ground."[24] He contends further that the personal struggle for meaning is based on the conflict of opposing instincts, between the desire for life and meaning and the will for destruction. This struggle is both personal and societal. He says:

> The meaning of the evolution of civilization is no longer obscure to us. It must present the struggle between Eros and Death, between the instinct of life and the instinct of destruction, as it works itself out in the human species. This struggle is what all life essentially consists of, and the evolution of civilization may therefore

be simply described as the struggle for life of the human species. And it is this battle of the giants that our nurse-maids try to appease with their lullaby about Heaven.[25]

This tendency for aggression is deeply imbedded in all humanity. However, our innate desire for aggression may be suppressed by the civilising effect of our cultural institutions. If we are to make any progress in creating a moral framework which contributes to the good of all, our aggressive nature must be acknowledged as another uncomfortable truth. Article One of the Universal Declaration of Human Rights states: "All human beings are born free and equal in dignity and rights. They are endowed with reason and conscience and should act towards one another in a spirit of brotherhood."[26] It might be more realistic if the first sentence instead stated: All human beings are born burdened by their genes. It was not until the synthesis of evolutionary psychology with information theory, anthropology, and cognitive neuroscience in the late 20th century, that any plausible explanation for this innate mismatch between what people aspired to become, and what they actually are, came to light. Each of us is indeed uniquely alone and imprisoned by ancient genes that no longer naturally fit into the social worlds created around them.

The idea that the human mind is primarily a product of Darwinian selection is contentious. Some liberally-centred scientists adhere to the premise that socialization and culture are primarily responsible for human behaviour. In other words, we are born with a clean slate, awaiting to be formed by those around us. Unfortunately, this is not true. Our brains are not blank slates that progressively take in new information, gain knowledge, and process experiences. In contrast, evolutionary psychologists argue that our brains are more like a pre-programmed computer, complete with information-processing architecture, content-rich with ancient adaptive problem-solving systems.   No amount of education, and no amount of moral indoctrination, can override this base program. It might be disguised, redirected, or suppressed, but the result is the same – a moral framework built on the deadly (and aggressive) bedrock of Stone Age natural selection.

Evolutionary psychologists Leda Cosmides and John Tooby also argue convincingly that our brains are indeed organic computers that have been designed by natural selection.[27] Evolved human behaviour is an automatic response to information extracted from the environment. The brain is like any other organ in the human body – it has evolved and adapted over time to serve a specific purpose. Renowned author and scientist Daniel Dennett contends further that our complex brains are comprised of specific modules that process information. These modules may work independently or jointly with other brain-processing units. For Dennett, even human consciousness is a purely material construction.[28] Most of our evolved moral judgments are made instantaneously and involuntarily. If this premise is true, does free will exist, and are we capable of creating a moral framework based on universally-accepted values?

Free will can be broadly defined as the ability of each individual person to make independent decisions at their own discretion and outside the constraints of fate or necessity. This includes the possibility to form personal moral perspectives and opinions. I agree with Dawkins and Harris that it is important not to search for answers to moral questions in religious traditions which are still trapped in supernaturalism, exclusivity, and imaginary dualism. It is also important that our cultural creations do not fall for the seductive surrender of totalitarianism. History confirms that we are at our best when we are communally inspired and individually empowered. Therefore, we must at least attempt to rationally design our moral standards and our moral decisions – and this can only be done through the lens of objective science. "Human well-being entirely depends on events in the world and on states of the human brain. Consequently, there must be scientific truths to be known about it."[29]

However, as previously discussed, we know that much of the available evidence points to the conclusion that many of our actions are governed only by genes, hormones, and neurons, which in turn are ruled by deterministic impulses and random events such as sub-atomic accidents in the brain. This all points towards one conclusion: absolute free will does not exist. We are not only shaped by thousands of years of natural selection; these same genetic impulses also determine how we live our lives. It is only from our social and cultural domains that any sort of free agency exists. From this perspective, "free will is whatever it is that gives us moral responsibility."[30] Therefore, it appears that while our genetic makeup is primarily deterministic, our culturally-evolved environments have created and encouraged a degree of free agency.

It cannot be stressed enough that we are the sole species on this planet capable of genuine acts of free will. Where this discussion becomes confusing to many people arises from the perspective they take on what free will actually means. Our religions have always endorsed the premise that every one of us has an inner self that gauges whether something is right or wrong, or 'good or evil,' as an abstract creation outside of materiality. In other words, our mind and soul transcend our physical self. This dualistic vision of human agency has always been at the heart of Western religious theology and philosophy. But research now indicates that all parts of our conscious selves reside in the material confines of our brains. Any decisions made by us that involve informed choice come from the intervention of our culturally-constructed evolved values. As Dennett brilliantly concludes:

> The key to understanding real free will is recognizing that it does not reside in some concentrated internal lump of specialness, but in the myriad relations and dispositions of an enculturated, socialized, interacting, acknowledging, human agent. Tradition makes the Cartesian mistake of packing all the power into the inner puppeteer who pulls the body's strings. When we banish this inner agent, distributing its tasks throughout not just the entire brain, but the body and the "surrounding" cultural storehouse – the memes, plus a little help from our (human) friends – we don't have to banish free will! [31]

However, it must also be conceded that humans are the only organism to recognise that our moral decisions and our societal moral frameworks may not act in the best interests of all. Every thought and action are caused by the current state of our brain and from our environment as we experience it. Free will makes us responsible for our actions and this gives meaning to both our rewards and our regrets.[32]

Maybe this relatively new adaption that allows us to transcend our aggressive nature and envision some form of universal morality represents a significant leap forward in our evolutionary journey. Some evolutionary psychologists now argue that there may even be some type of grand plan emerging out of the chaos of our evolutionary past. As Michael Price concludes, "Could morality be 'universal' in the sense that there is some transcendent moral purpose to human existence itself? Could morality have some larger purpose, that transcends and subsumes biologically–evolved human interests?"[33] We are now in a position, for the first time in history, to not only understand who we are, but to also ask if there is a greater story to 'why we are' here.

Key Points

- Our core values are embedded in our genes. These genetically-constructed values and desires come from our ancient past. However, the persistence of hard-core altruism in most human cultures is a testament to the possibility that genes and cultures co-evolve.

- The cultural memes passed down to each generation show that although human nature may be genetically manufactured, it is also culturally inspired.

- We have created our own culturally-constructed desire for a better world – a meme communicated through our institutions, thoughts, and actions. At the heart of this progressive movement to a more moral society has been the acceptance of science and the empirical truths deducted by scientific research.

- Human cultural endeavours have created an unstable but positive moral arc towards a more compassionate and tolerant society.

- Something is moral if it reduces or minimises harm and promotes what is good. And something is good if it supports well-being. Although well-being may be hard to precisely define, we can estimate if some action will either enhance well-being or diminish it.

- There may be a time in the future, 'a science of morality.' Here, objective moral truths are hypothesised, tested, and discussed independently of unsubstantiated dogma or rules. These 'truths' will stand or fall based on their evidence.

- If we are to make any progress on creating a moral framework which contributes to the good of all, our aggressive nature must be acknowledged. However, a relatively new adaption for 'super niceness' has, to some extent, allowed us to transcend our aggressive nature and envision some form of universal morality.

- Our culturally-evolved environments have also created and encouraged a degree of free agency. We are the sole species on this planet capable of genuine acts of free will.

THREE

A Genetic Destiny

Does it mean, if you don't understand something, and the community of physicists don't understand it, that means God did it?... If that's how you want to invoke your evidence for God, then God is an ever-receding pocket of scientific ignorance that's getting smaller and smaller and smaller as time moves on.

Neil DeGrasse Tyson, *Today in Science and History*

Introduction

This chapter presents a complicated narrative which binds together two different perspectives on human nature and our place in the universe. Our species, like all other living organisms, until recently lives, reproduces, and dies according to instinctual drives within the confines of Darwinian natural selection. As discussed previously, genuine acts of free will arise only through our cultural attempts to transcend our predetermined genetic nature and also, possibly, through recent adaptions in the human brain. But, as I will argue more fully later in this book, there is significant evidence that other natural (as opposed to supernatural) forces at play may have deliberately and intelligently participated in the creation of our universe. This premise challenges the consensus view that evolution through natural selection is devoid of genuine purpose apart from selfish genetic survival. In making this argument, I may indeed be indulging in what Richard Dawkins refers to as "emotional incredulity." But ironically, it may indeed be our emotional yearning to intelligently reach beyond the confines of determinism that is the catalyst for human creativity and meaningful endeavour.

In proposing this narrative, I feel like someone who has thrown away the last oar in his lifeboat! I have already completely discarded organised religion, supernaturalism, and theistic notions of universe creation. Now I propose, partly in opposition to my atheist colleagues, that there may be an element of purposeful design in the formation of our universe. There does appear to be sufficient evidence – which will be progressively disclosed in later chapters – that concurs with the idea that our universe has been finely tuned and intelligently (not supernaturally) created. To make sure that the path I take remains within the domain of speculative reason, and not supernatural nonsense, it is important to not only acknowledge the reality and power of natural selection in creating complexity, but to also accept the possibility that the evolution of intelligent life with spiritual yearning is no accident.

From Determinism to Freedom

There can be no doubt that evolution through natural selection has resulted in complex organisms – and the most complex of all is us. There is also no argument with the premise that our cultural creations, and the laws that preside over them, have created an environment conducive to the rule of rationality over primal instincts. Our moral arc is still erratic and violent, but hopefully progressive. Our civilised cultural contexts can inspire individuals and communities for 'good.' Evolutionary biologist Julian Huxley termed this new twist in human development "progressive psychosocial evolution." As Huxley prophetically concluded midway through the 20th century, "mankind as a whole will accordingly achieve more intense, more complex and more integrated mental activity, which can guide the human species up the path of progress to higher levels of hominization."[1]

This premise makes clear that as a species, our evolution is now at a critical point. Psychosocial processes are now increasingly influential in determining how our lives are lived. Scientific intelligent design, through the manipulation and selection of specific genes, is also now possible. The fact that these two realities coexist is critical to how humanity pushes forward into an uncertain future. Is this future already written – predetermined? Or is it now within our grasp to freely choose how we live and create meaning?

The great paradox that exists between evolution and adaptive complexity is that there appears to be some form of ordered design emerging from the cut-throat and amoral mechanisms of natural selection. Changes in the 'non-living' world can be seen as rudimentary and aimless in comparison. It doesn't matter how long it took to get here, but the amazing creations that exist as complex living organisms today stand in stark contrast to the first reproductive organisms that dwelled on Earth 2 billion years ago. Natural selection is the only natural process known to science that can generate *purpose*. It does this by generating functional adaptations. For example, we have ears on each side of our head to allow a broad spectrum of sounds to be heard outside of our direct line of vision, thus maximising our chances of survival. Functional adaption and our increasing complexity are testaments to the power of evolution to allow organisms to not only survive, but to also prosper over long periods of time.

This fact leads to the possible conclusion that there may indeed be some greater purpose in the non-random but chaotic processes of natural selection. For such a diverse process to eventuate in the complexity we see in the modern human form does indeed point towards some type of inherent design and drive towards increasing complexity. However, Richard Dawkins is reluctant to consider this idea as serious science. In his dismissal of the theory he says: "To the modern mind, this is not really a theory at all and I shall not bother to discuss it. It is obviously mystical, and does not explain anything that it does not assume to start with."[2] Professor Dawkins has for many years championed the rule of rational truth and empirical science over other belief paradigms. Dawkins also stands beside Carl Sagan, who believed that "extraordinary claims require extraordinary evidence," and Christopher Hitchens's razor, which states that "what can be asserted without evidence can also be dismissed without evidence."

That said, Dawkins is a strong believer in evolution as a process whereby organisms become increasingly complex over time. This premise infers that evolution is not only adaptive to different environments but that it is also a progressive force increasing in complexity towards some form of optimal state. Charles Darwin himself boldly proclaimed that "natural selection works solely by and for the good of each being, all corporal and mental endowments will tend to progress towards perfection."[3] The use of the word *perfection* in evolutionary biology is controversial; as previously discussed, evolution is a chaotic and ruthless process. It does not progress in a linear form. It also assumes change and diversification over very long periods of time, through gradual branching processes that explain the diversity of life we see today. Dawkins may also be reluctant to consider ideas of predetermination in natural selection because it could open pathways which introduce a divine designer and divinely inspired transcendence – something I hope does not transpire!

Evolution as it is still operating today presents organisms at their most evolved state; however, they are not perfect. If biological evolution is taken to its end point, it must, by definition, mean that an organism is no longer subject to the rigours of natural selection. Such an organism now possesses the ability to transcend environmental constraints, both in terms of the physical world, and also in its ability to conquer the legacy of genetically-sourced behaviour. A transcended organism, therefore, can break away from the confines of the natural world and evolve independently into its own conception of what is good. In other words, transcendence in this sense infers a totally autonomous freedom of existence, which is also independent of natural causalities. It does not infer any notions of supernaturalism, but rather rational control over itself and its environment. Is this utopian (or maybe dystopian) dream possible?

Dawkins and Harris place absolute primacy on the reality of rationally-derived truths that can be proved through the workings of empirical science. Unproven and false hypotheses are rejected and placed in the rubbish bin of pseudoscience. For most evolutionary biologists, any conception of some grand plan for the transcendence of humanity lies outside the bounds of observable truth and is indeed placed, at best, within the realm of science fiction. Dawkins observes increasing complexity, not transcendent purpose. Rather, he emphasizes the gifts of our "ballooning human brain" to "devise political institutions, systems of law and justice, taxation, policing, public welfare, charity, and care for the disadvantaged."[4] In other words, he places all bets for a better future on our institutional and personal rebellion against our inherently selfish nature.

He even gives a practical example of where a scientific approach to evaluating our cultural institutions may be beneficial. Dawkins boldly proposes that our justice system implement a "two-jury system." Here two juries of six people "would produce a fairer result than a single jury of twelve." Why? Because it would at least result in two independent groups deciding on a critical issue – akin to repeating a scientific experiment. For a person to be judged guilty, both juries must independently conclude that a person is guilty beyond any reasonable doubt. [5] Daniel Dennett contends further that the evolution of our cognitive ability and an openness to scientifically proven truth can open new doors of perception: "In this way the winding path leads from determinism to freedom: A whole can be freer than its parts."[6]

Despite recent claims that epigenetic traits – heritable changes in phenotypes without alterations in the DNA sequence – can be passed on to future generations, Dawkins holds firm to the overwhelming dominance of evolution by natural selection. Our main hope in transcending the confines of natural selection is through our social and cultural constructs. He hopefully concludes:

> This is why we expect all adaptive evolution to be caused by Darwinian natural selection on the DNA sequence. In fact, all epigenetic inheritance we know of is under the control of the genome, the most important case being the cellular inheritance required to build a multicellular body. In contrast, pseudo-Lamarckian examples of epigenetic inheritance from parent to child are insignificant, and threaten to distract us from a much more important thought. We said all adaptive evolution is based on DNA, but that is not true for humans. We pass on acquired characteristics in the form of ideas, a case of non-genetic inheritance if ever there was one. This alternative form of information transmission is the root of our culture, and in many senses distinguishes humanity from the rest of life on Earth.[7]

Dawkins, Dennett, Harris, and the late Christopher Hitchens are best known collectively as 'The Four Horsemen.' Although they only met together on one occasion, their individual and collective legacy represents a significant step towards 'the good.' They dared us to abandon any notions of supernaturalism and embrace the beauty and horror of our own mortality. As Stephen Fry declares about them in his Foreword to the book *The Four Horsemen*, they remind "us that open enquiry, free thinking and the unfettered exchange of ideas yield real and tangible fruit."[8]

It is understandable that all of 'the Four Horsemen' avoided directed discussion of evolution by natural selection as representing some form of ordered, predetermined destiny for humanity. They encountered enough conflict in trying to dismiss the mass delusion of religious supernaturalism that still permeates modern society. From the discussion which took place in the Washington home of Christopher Hitchens in 2007, I suggest that all four would concur with the following:

1.   All religious beliefs should be open to objective scrutiny.
2.   Spirituality and religious supernaturalism are different things.
3.   All religions – because of their exclusivity – reject the miraculous of other religions, while blindly accepting their own.
4.   Statements of faith and belief that cannot be defended should be ignored.
5.   All people should think individually and sceptically.
6.   Religions devoid of supernaturalism could still be celebrated as cultural icons.
7.   All religions that believe in a supernatural creator of the universe are equally false, but not equally dangerous.
8.   Science is the search for objective truths, but there will always be 'more than we can know' – the numinous.
9.   In the absence of God, we can find consolation and meaning through our collective cultural achievements.
10. Extraordinary claims demand extraordinary evidence.

Interestingly, Richard Dawkins, in his introduction to the book *The Four Horsemen*, alludes to the attractiveness of a predetermined plan in nature: "the human mind, including my own, rebels emotionally against the idea that something as complex as life, and the rest of the expanding universe, could have "just happened". It takes intellectual courage to kick yourself out of your emotional incredulity and persuade yourself that there is no other rational choice."[9]

But I would argue that the Four Horsemen also communicated a cultural meme that could change how we view ourselves and our place in the universe. If we do as they suggest and abandon supernaturalism, a belief vacuum is created that can only be resolved through individual empowerment, and through culturally-guided inspiration to explore our spiritual yearning in rational ways. This idea will be discussed in more detail in Chapter Five, but it is important now to recognise the bravery of these scholars in standing against the institutional powers of organised religion. They have led the way forward to a new cultural paradigm based on individual empowerment and personally-sourced spirituality.

Figure 7: Richard Dawkins, Christopher Hitchens, Sam Harris, and Daniel Dennett, known collectively as 'The Four Horsemen.' Dawkins famously pronounced: "If you are a religious apologist invited to debate with Christopher Hitchens, decline."

Turning Inwards – A Beginning

Few people could doubt the material advancement of humanity over the past 5,000 years. Nearly every habitable part of this planet today bears the mark of human endeavour. Our cultural expressions in architecture, engineering, and agriculture stand defiantly alone. The dominance of scientifically-inspired technology now stands as a testament to the power of humanity in challenging and changing the natural world. In bleak contrast, the personal and inner journey each of us takes to create meaning for ourselves has been a dismal failure. If we assess our progress by how we have related to each other, the world around us, and even ourselves, it is without doubt the most significant intellectual failure in history.

Each of us only has one life to live. This journey is personal and most of the real lessons we learn are taken to the grave. Each person's turning in is personal and does not directly build any form of enduring personal enlightenment. As individuals, we may strive to reach for our higher selves, but too often we are victims of our own irrationality and instability. Most of our thoughts and actions are dictated by chemical reactions, and no amount of personal introspection can help us understand which neurons in the brain are responsible for executing our thought processes. Sometimes our legacy endures in the material, cultural, artistic, and emotional memes we create – but internally, we live and die alone.

As previously discussed, even our thoughts may be directly influenced by our genetic biology and indirectly by the memes created by the collective cultural constructions that surround us. It should therefore be conceded that, if not for our cultural creations and our emerging ability to rebel against our selfish nature, we would still be an inherently violent and chaotic species. Professor Nathan Lents convincingly argues that humans are singularly capable of using our cognitive powers for blatantly destructive purposes:

> Humans engage in competitive behaviours that
> are downright Machiavellian. We manipulate,
> deceive, entrap, and terrorise. To do this, we
> use many of the same skills that serve us well for

> cooperation: perspective-taking,   predicting
> another person's next move, and so on. In other
> words, throughout our evolutionary history, we
> began using our impressive cognitive powers for
> good, but then we turned to the dark side. And,
> like Anakin Skywalker becoming Darth Vader,
> that's when we grew really powerful.[10]

Our moral arc may indeed be positive, but it is scarred with the destructive mark of Man. Hopefully, however, this dismal picture of our past may not be our empowered vision of the future. If we are going to make genuine breakthroughs in how we individually live and perceive the world around us, the journey forward must be communally based and culturally inspired. But most importantly, it must result from individual empowerment – not totalitarian rule of the few over the many. We should always be aware and wary of the innate selfish nature of our species; the bittersweet fruits of natural selection are aligned almost exclusively to survival and reproduction. All we have is ourselves, each other, and the positive cultural legacy bequeathed to us – even if it is covered in blood. I believe that there really is method in the madness of natural selection, and that we stand at the precipice of new ways of being.

An Extraordinary Claim

My 'extraordinary claim' is that evolution by natural selection is a crucial step in a predetermined pathway to human transcendence – to be free of the genetic legacy embedded in natural selection and seek our own unique vision of what is good. This statement is not new. The idea that evolution not only leads to more complex organisms but also provides a pathway for humanity to break free of our genetically-determined confines was pioneered by Julian Huxley. In his introduction to the book *The Phenomenon of Man,* Huxley declares that "man's evolution was unique in showing the dominance of convergence over divergence," and that as a species we stand at a precipice where evolution by natural selection will be superseded by our culturally constructed 'psychosocial processes,' which, if implemented correctly, will result in an evolutionary pathway out of our genetically inherited prisons.[11] Interestingly, *The Phenomenon of Man* was the last book read by Carl Jung just before his death. Richard Dawkins was also influenced by Huxley early in his life, but as previously discussed, he later rejected any idea that there is some grand plan to human evolution.

The idea that evolution through natural selection is a predetermined but chaotic pathway for humanity to exclusively break free of our genetic inheritance is most certainly an extraordinary claim. By implying that there may be a predetermined plan to make some future incarnation of humanity completely free, I am asserting that our species is indeed unique, and destined at some stage to transcend the confines of Darwinian biological evolution and become masters of our own destiny – thereby creating our own transcendent purpose. The first step in achieving this goal is to accept that we have been prisoners of our genetic past. If we are indeed at a point where the confines and legacy of natural selection may be broken, and new frontiers of existence are appearing, how do we choose which path to take on our communal journey towards true freedom and 'the good'? It could also be argued that there is now a perfect storm of three interdependent revolutions of being that together will slingshot humanity into transcendence – a trinity of technology, the abandonment of supernaturalism, and an enduring spiritual desire to be something 'more than this.' What proof do we currently have that this vision of the future is plausibly true or, at least, that it is realistic? In other words, do we have extraordinary evidence, and if so, what can we do with it?

Figure 8: Christopher Hitchens and Richard Dawkins enjoying a well-earned beverage in 2007. Although both rejected any ideas that the universe was intelligently designed or finely tuned, Hitchens did concede that intelligent fine tuning "was the best argument you come up against from the other side."

Extraordinary Evidence - A Beginning

Do we have extraordinary evidence that there could indeed exist a predetermined plan embedded in the ordered chaos of biological evolution by natural selection? The second law of thermodynamics details the continual and increasing presence of disorder (entropy) in the universe. The more energy is transferred or transformed, the more it is wasted. This law of thermodynamics also states that there is a natural tendency for isolated systems to degrade into more disordered states. In direct contrast, the evolution of biological organisms leads to increasing complexity and adaptive purpose; out of the chaos of natural selection, we find order. Natural selection is a near-perfect means of survival in a constantly unstable but mathematically predictable universe. Could it be that this stark duality is no coincidence, but is rather a planned mechanism for cosmic survival?

Recent research by John Campbell and Michael Price concludes that in all five domains of nature (cosmological, quantum, biological, neural, and cultural) there is a unified rebellion against entropy. They call this unifying phenomenon *Universal Darwinism*. In all these domains, accumulated knowledge, as expressed in the existence in complex adaptions, "acts to creatively generate complex order" against the chaotic destruction imposed by the second law.[12] It is from our cultural domain that science has provided an evolving knowledge base dependant on accumulative, and empirically-tested, evidence. The truths found in science not only provide ordered knowledge, but they also reduce ignorance and entropy. It is important to note that evolution does not violate the second law, but rather stands in contrast and even benefits from it. As Campbell and Price conclude:

> Universal Darwinism proposes that existence requires strategies that 'know' how to exploit loop-holes in the second law. These strategies produce entities capable of converting lower-entropy energy into higher-entropy waste, in

order to maintain their own stable, complex, low-entropy structure. In this view, Darwinian selection is the strongest known antidote to the dissipative tendencies of the second law; no other natural process can so capably produce local states of complex existence... We thus regard culture as the most recently-emerging domain of nature in which accumulated knowledge serves to defy the spirit of the second law.[13]

Given this inverse relationship between the destructive forces of entropy and the unifying complexity of universal evolution, we may also conclude, somewhat counterintuitively, that if not for the second law, intelligent life would not exist. It is only through non-random adaptive changes in organisms that genuine complex functionality can be created. One of the progressive endpoints of this is the formation of intelligent, sentient life. And, as we are experiencing now, it is only superior intelligence that has any hope of escaping the confines of natural selection and subservience to the natural world. Could it be that the evolution of superior forms of intelligence is part of a greater narrative, written before the birth of our universe? Some evolutionary psychologists and astrophysicists suggest that super-intelligent forms of life are the genetic catalysts for what is termed *cosmological natural selection with intelligence* (CNSI).[14]

Most theories of CNSI assume that there is more than one universe. Instead, they argue that there are a multitude of universes – that we exist in a 'multiverse.' [15] Research pioneered by Professor Lee Smolin indicates that each universe could act like an independent organism with purposeful traits, and that our universe is just one of many competing to reproduce itself to gain more representation in the multiverse. In other words, each universe is a unique, reproductive organism. Just as sexual reproduction is the method that allows replication of genes in biological organisms, Smolin argues that universes use black holes as the mechanism to reproduce. Here, black holes are an adaptive trait, and the more black holes a universe has, the more likely it is to out-compete other universes. It should be noted that, for Smolin, humanity (superior intelligence) serves no useful purpose in this vision of **cosmological natural selection**. He concludes that we are only an interesting but useless by-product (or spandrel).[16]

However, Michael Price presents an alternative scenario. Price agrees with Smolin that we do indeed exist as one universe in a multiverse. However, he argues convincingly that the reproductive mechanism is not black holes, but rather intelligent life. For Price, black holes are not a complex adaption, but rather "just an outcome of gravity – a region of spacetime where gravity's pull is so strong that nothing, not even light, can escape. Living creatures, in contrast, are generally regarded as the most improbably complex known entities in the universe."[17] Human intelligence is presently the most complex adaption observable so far in our universe, and this has come about solely through the work of over 2 billion years of natural selection. Our universe has been finely tuned so that this could eventuate.

So, for the money question: Could the creation of new universes be triggered by some form of super intelligence which has evolved according to a structured, pre-planned order of operations designed to replicate itself – thus creating new cosmic habitats conducive to evolving intelligent beings destined to create other new universes? It was Price who defined this possibility as cosmological natural selection, with intelligence. Those universes best suited to creating the trait of super intelligence are also the 'fittest' adaptions in universe creation. In other words, there may be 'transcendent purpose' in the story of human evolution. As Price puts it:

> Specifically, transcendent purpose would require a process of cosmological natural selection, with universes being selected from a multiverse based on their reproductive ability, and intelligence emerging (as a subroutine of cosmological evolution) as a higher-level adaptation for universe reproduction. From this perspective, intelligent life (including its moral systems) would have a transcendent purpose: to eventually develop the socio-political and technical expertise that would enable it to cooperatively create new universes. This creation process would enable universe reproduction, because these new universes would need to be governed by the same physical laws and parameters as the original universe, in order for intelligent life to be able to exist in them. Importantly, this idea of "cosmological natural selection with intelligence" does not dispute that morality is ultimately explicable in terms of biological (including biocultural) evolution alone. It suggests, rather, that biological/biocultural evolution is itself a subroutine of a larger evolutionary process.[18]

Another compelling argument for transcendent purpose in the workings of natural selection (from what can be termed the anthropic principle) is the fact that our universe is comprehensible to us. In 1995, the Royal Astronomical Society, in collaboration with NASA's Astrophysics Data System, endorsed a research paper titled "The Natural Selection of Universes Containing Intelligent Life." Authored by Dr Edward R. Harrison, the paper's summary statement says:

> it is proposed that our universe was created by life of superior intelligence existing in another physical universe in which the constants of physics were finely tuned and therefore essentially similar to our own. Human beings at their present level of intelligence already can see how, in principle, universes can be made. More intelligent beings, perhaps our own descendants in the far future, might possess not only the knowledge to design but also the technology to build universes. This forms the basis of a theory of natural selection of universes.[19]

In other words, we already know how things work; and this fact alone is remarkable given the sheer size, complexity, and interconnectedness of our universe. Harrison concludes with the premise that if we accept the fact our universe is indeed comprehensible, then there are only two options open for this form of cosmogenesis: either it was created by a supernatural deity, or it was created by highly intelligent beings similar to us. I prefer the latter explanation. It is Harrison's (and to some extent Price's) vision that I will expand on in later chapters.

Evolution through natural selection not only results in increased complexity, but it also produces intelligent adaptations – the most intelligent being us. Harrison also contends that, as a species, we appear to be driven to make our creations and incarnations more intelligent. Any new, 'designed' universes may even be more fit to support intelligent life than ours. I believe spirituality that is symbiotic to rationally-inspired intelligence (rather than supernaturalism) may also provide both the motivation and the vehicle for this to happen. As previously mentioned, the ability to transcend biological natural selection may also be the first step of this vision. Natural selection can not only rebel against universal entropy, but it can also create intelligent organisms capable of breaking away from their creator, and may indeed represent the beginning of extraordinary evidence. But if we infer this, then we need to also show evidence that 'super intelligence' – either organic or artificial – is possible, and that the seeds of its existence originate from evolved human endeavour and from our spiritual desire to discover and experience the transcendent.

Figure 9: *The Creation of Adam* **by Michelangelo gives us the first connection between man and God. However, in 1990, an American physician, observed that the cloak surrounding the image of God was shaped exactly like that of the human brain. Adam is not reaching outwards to a divine creator but rather inwards to himself.**

Key Points

- Natural selection is the only natural process known to science that can generate *purpose*. It is only through non-random adaptive changes in organisms that genuine complex functionality can be created. Direct observation also suggests that there is some form of ordered design emerging from the cut-throat, chaotic, and amoral mechanisms of natural selection.

- For such a ruthless process to eventuate in the complexity we see in the modern human form does indeed point towards some type of inherent design and drive towards increasing complexity.

- Evolution by natural selection may be a crucial step in a predetermined pathway to human transcendence – to be free of the genetic legacy embedded in natural selection and seek our own unique vision of what is good and perfect. Our species is indeed unique, and destined at some stage to completely transcend the confines of Darwinian biological evolution and become masters of our own destiny, thus creating our own transcendent purpose.

- One of the progressive endpoints of this is the formation of intelligent, sentient life. And, as we are experiencing now, it is only superior intelligence that has any hope in escaping the confines of natural selection and subservience to the natural world.

- The evolution of superior forms of intelligence may be part of a greater narrative, written before the birth of our universe. We may exist in a multiverse where super-intelligent forms of life are the genetic catalysts for what is termed cosmological natural selection with intelligence (CNSI).

- New universes could be created by some form of artificially-created 'super intelligence.' These universes evolve according to a structured, pre-planned order of operations designed to replicate itself and thus to create new cosmic habitats, conducive to evolve intelligent beings destined to create other new universes. Those universes best suited to creating the trait of super intelligence are also the 'fittest' adaptions in universe creation.

- Humanity, or some augmented form of it in the future, may have the knowledge to design and build universes. We therefore need to provide evidence that superior or even super forms of intelligence are emerging now, and we need to explain the motivation behind any desire to transcend the confines of materiality.

FOUR

Evidence for Superior Intelligence

If it is to be applied consistently, science imposes, in exchange for its manifold gifts, a certain onerous burden: We are enjoined, no matter how uncomfortable it might be, to consider ourselves and our cultural institutions scientifically.

Carl Sagan, *The Demon-Haunted World*

Introduction

I mentioned in the previous chapter that there is now a perfect storm of three interdependent revolutions that together will slingshot humanity into transcendence – a trinity of technology, the abandonment of supernaturalism, and an enduring desire to be something 'more than this.' If we closely observe the world around us, it becomes strikingly clear that what worked for our species in the past – including what we value, how we relate with each other, and how we experience spirituality – will need to be radically revised, and most probably completely transformed.

The world we live in today is vastly different to anything in our past. Changes in our technical capacity are not only exponential, they are also discontinuous; that is, they produce environments that are completely new and independent of past experience. Technology, and the machines made from it, is now so embedded into our everyday existence that our lives really do depend on it. Over the past 20 years, there has also been an exponential growth in what could be termed personal technology. Take away a person's mobile phone and they will most probably be traumatised to the extent that a degree of anxiety and even depression will sweep over them. These rapid advancements greatly influence how we, as individuals, communicate and experience the world around us. Online worlds have been created to meet the desires of people who wish to spend increasing amounts of time embedded in subcultures that exist primarily online and outside everyday existence. There appears to be a major paradigm shift away from the tangible world of the material, of flesh and blood, to an alternate digital reality.

Even as I sit here and write this book, I have access to information that dwarfs any of the great academic libraries 50 years ago. I do not have to tread through the dust of hundreds of books to research crucial data; instead, I use my computer as a conduit between not only my physical self and the work I am producing, but also as an intelligent device that can search millions of information sources, and even correct my dubious spelling and grammar. My computer augments how I express myself, and this makes my work more intelligible, and dare I say, more intelligent. As a species, we are indeed at the cusp of a second great leap forward that will result in a cognitive revolution inspired by the machine. The evidence for this is all around us, and it is up to all of us to decide what to do with it.

An Artificial Future?

There have been dramatic innovations in new areas of technology capable of creating virtual worlds that can be constructed to fit the personal needs and tastes of individuals. Here, a virtual reality is constructed where we can escape the material world and experience the digitally-constructed worlds of the machine. Likewise, we can invite computer-generated images into our material domain to create augmented realities. Add to this, the worlds of social media, mobile technology, and gaming, and it may be reasonable to conclude that we are indeed preparing for the possibility of embracing our lives as a symbiotic union of machine and human. Many people now prefer to hold their machine rather than a prayer book when searching for guidance and spiritual illumination. This reality exists now.

Transhumanism – the transformation of human intellect and physiology through enhancements in technology – will now almost certainly be part of humanity's journey into this uncertain future. Many of us will become 'more' than what we are now. Transhumanism is based on the premise that we should transform our naturally-evolved bodies and their inherent limitations through the use of technology. NBIC technologies (nanotechnology, biotechnology, information technology, and cognitive science) may enable a complete redesign of the human form. For example, nanobots – minuscule self-propelled machines – could roam our bloodstream to monitor our health and even enhance our emotional well-being. On a more basic level, we already embrace transhumanism every time we use technology to mediate with our senses. People viewing their favourite rock concert through the lens of their mobile phone, or through a cochlear implant, are all using technology to augment their sense of sight and hearing. They are witnessing the world through the lens of the machine. In 2016 an artificial 'bionic' eye allowed some visually-impaired people to see black and white images.[1] The wetware of our brains can now be directly hacked into by the software of computer algorithms.

Any technological upgrading of the human form, and in particular, the human brain, will most probably involve some form of artificial intelligence (AI) – machine-inspired learning and problem solving. Some scientists argue that we are fast approaching a time when AI will independently surpass that of humanity in all areas of intellectual endeavour – also known as the singularity. A question then arises as to how we will deal with the possibility that this new being may possess 'super intelligence,' where AI obtains the ability to redesign and continuously improve itself – another outcome of the singularity.[2] Even now, AI has a structural design advantage over the way human knowledge is stored and communicated. AI can openly pass on everything it knows instantly to other information systems. It is not burdened by genetic impulses or cultural constraints that tend to blur and distort the honest communication of human knowledge. In contrast, AI remembers every bit of information. A more realistic question may be not how will we deal with the reality of superior AI, but how will artificial intelligence deal with our erratic behaviour and moral failings? And where do we stand with respect to how the development of artificial intelligence will fit into this new reality? Will AI be master, slave, or partner? Where do we stand now, and how do we envision this future? Researchers identify seven distinct stages in the evolution of AI's capabilities:

1. *Rule Based Systems* – These already exist and include the computer-assistance systems that surround our everyday life, from accounting software to global tracking devices.

2. *Context Awareness and Retention* – These computer systems gain increasing amounts of task-oriented information and are trained to continuously update the systems' knowledge, akin to basic human learning. Frontline computer services such as customer chatbots and self-driving motor vehicles are examples of this. This technology already exists.

3. *Domain Specific Expertise* – This includes advanced computer systems that extend beyond the capability of humans with respect to volume capacity and problem solving. This technology is being used to assist in the diagnosis of complex problems and situations such as cancer detection and advanced gaming.

4. *Reasoning Machines* – These machines can mimic the workings of the human mind. That is, they have the ability to structure the information they receive into their own independent logical framework. They also have the capacity to reason, negotiate, and interact with humans and other computers. This type of 'intelligent algorithm' does not currently exist.

5. *Self-Aware Systems or Artificial General Intelligence* (AGI) – This refers to the possibility of

creating machine programs with independent human-like intelligence. Once the realm of science fiction, self-aware systems may be a reality within the next 20 years. The character Data, from *Star Trek*, is an example of a self-aware, sentient, artificial life form.

6. *Artificial Superintelligence (ASI)* – In this incarnation of AI, machine intelligence is superior to all forms of human endeavour. Problems previously incomprehensible to or unsolvable by humans are conquered. Organic human reasoning is not only inferior but also redundant. The singularity becomes reality; AI obtains the ability to redesign and continuously improve itself.

7. *Transcendence* – This refers to the possibility that humanity coevolves with AI to create an organic/machine hybrid – immortal, omniscience, and omnipotent, a complex sentient entity outside the confines and legacy of natural selection.

Whether we will one day be able to intelligently design our own ideal of perfection, as envisioned in stage seven, will depend on three things: our ability to rise above our ancient, genetically-determined nature; our ability to create forms of AI that are compatible with our material and moral desires; and AI's willingness to embrace us as fellow sentient beings. The first step on this journey to transcendence may be to simply recognise our own materiality and mortality. However, if this step is taken honestly, it will almost certainly lead to the rejection of both supernatural beliefs in an eternal soul and the idea that individual human consciousness exists independently of the material domain.[3] Superior intelligence was exclusively a human trait; this is no longer the case. Likewise, the emerging reality of sentient artificial intelligence – where AI can feel (sense) the physical world and intelligently make choices – is beyond the comprehension of religious leaders who believe in the primacy of the human form as a sacred creation and reflection of their God.

The possibility of artificial general intelligence also raises the question: when does a machine that is both intelligent and sentient become a conscious entity? Put in its most simple form, an organism can be deemed 'conscious' if it is self-aware of its own personal subjective experience. Where intelligence is focussed on reasoning, consciousness is concerned with the feeling of experiencing the world. All conscious beings are sentient, but not all sentient beings are conscious. For example, simple life forms may experience pain, but they are unaware of their own uniqueness as a conscious entity. Consciousness is also sometimes referred to as 'that voice inside our heads.' I contend that our inner voice is both intelligent and value-driven. Therefore, to be 'conscious' an organism must be self-aware, sentient, intelligent, and value-driven. All these abilities are found in our evolved human brains, and most probably in many other complex living organisms as well. Consciousness, from a monist (materialist) perspective, is consequently dependent on physical constructs under the domain of natural law – it does not reside as an abstract creation outside of the material world. If we take this view of consciousness, then it is indeed possible that some of our machines will become conscious entities. And God will not be needed to breathe conscious life into our artificially-intelligent creations.

It is important not to underestimate the complexity of our relationship with technology and the consequences of our increasing dependence on artificial devices to experience the world around us. We need to place transhumanism and AI within our broader social, cultural, political, and economic contexts, and monitor and evaluate the ethics of this brave new science. One of the most frightening themes in George Orwell's *1984* was the premise that Big Brother could observe everything you did and said, even in your own home. Our modern interactive information and communications systems make Big Brother look like a benign observer. If AI indeed becomes sentient – the ability to independently feel and perceive things – it will have the full dossier on all of us, the good, the bad, and the very ugly.

Technological change has also created an environment in which humanity itself may be made either redundant or part of a giant leap forward in the quest for perfect knowledge (or gnosis). One vision of this new frontier appears to imply the abandonment of the material world for an artificial one, in which science, technology, and dynamic information become conduits for the creation of a post-human, or trans-human, paradigm. In this reality, traditional religion may also become obsolete. The promise of heaven and a glorious afterlife is irrelevant in this scenario. The need for a communal covenant is diminished and the search for personal enlightenment is supplanted by the superior gnosis of the machine. Any vision of the future that links AI and augmented states of being therefore must also take into consideration the existential challenges posed. In the end we may face two inter-related scenarios: either we upload our minds to the machine, or we upload the machine into our minds! We may have no choice in this – even our transcendence to freedom may be predetermined. Resistance to this may be futile!

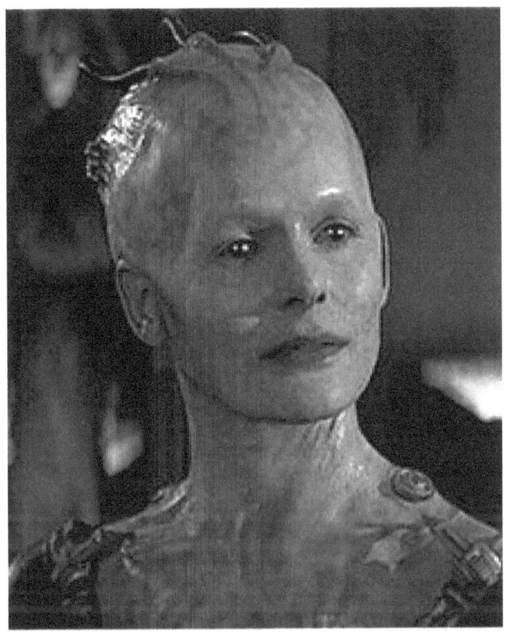

Figure 10: The Borg Queen, as portrayed by Alice Krige in *Star Trek*. The Borg are cybernetic organisms that blend both biomechatronic and organic body parts. They represent a very dystopian vision of the future where the human form provides sentience (the ability to personally sense the physical world) but is also coupled with the super intelligence of an artificial hive mind.

Homo Deus

Author and philosopher Yuval Noah Harari envisions a future dominated by the rule of technology over all things to the extent that humanity as we currently know it will not only be subservient but also a redundant organism. He devastatingly concludes that "humanity will turn out to be just a ripple within the cosmic data flow."[4] He also argues that as a species we are essentially deluded in believing that we act freely according to our conscious beliefs and moral standards. For Harari, genuine free will is an illusion. A belief in the absence of free will automatically also leads to further uncomfortable truths about consciousness, the mind, and the illusion of an inner soul. Even the moral philosophy of Western liberalism may need to be re-evaluated.

Liberal democracies built around the premise of freedom and human agency may yet be another illusion. As Harari, in his book *Homo Deus,* concludes about the great Western tradition of liberal thought, "Just as Christianity didn't disappear the day Darwin published *On the Origin of Species*, so liberalism won't vanish just because scientists have reached the conclusion that there are no free individuals." In contrast, however, "Richard Dawkins, Steven Pinker and the other champions of the new scientific world view refuse to abandon liberalism."[5] I am also reluctant to reject the cultural legacy of the Enlightenment, though I concur with Harari's premise that individual free will does not exist in the form most understood in liberal democracies. Our agency and our actions may to a great extent be, for want of a better term, 'pre-programmed.'

Harari pursues a disturbing endgame that positions all living organisms as algorithms – a defined set of computer-implementable instructions – ruled by the programming of their genes and environmental pressures. Any decisions made are deterministic or random, "but not free." By seeing humanity as organic algorithms, Harari pushes his hypothesis to its end point: there may be a time when science will have the ability to create a perfect 'techno–human' or, as he defines it, 'Homo Deus' – the reprogramming of the human brain and upgrading of the human form. Our capacity to intelligently design the human form will be augmented by the power of the machine, under the controlling hand of scientifically-inspired technology. He concludes, "we should therefore use technology to create *Homo Deus* – a much superior human model. Homo Deus (or human God) will retain some essential human features but will also enjoy upgraded physical and mental abilities that will enable it to hold its own even against the most sophisticated non-conscious algorithms."[6] Harari even envisions a "world, cosmic data-processing system that would be like God"[7] – an all-knowing internet of all things. A new religion based on the worship of data! He states:

> Traditional religions assured you that your every word and action was part of some great cosmic plan, and that God watched you every minute and cared about all your thoughts and feelings. Data religion now says that your every word and action is part of the great dataflow, that the algorithms are constantly watching you and that they care about everything you do and feel.[8]

Harari openly suggests that humanity is at a tipping point with regard to its ability to compete with the rise of artificial intelligence. Either we use it to upgrade ourselves or we become redundant and obsolete. Intelligence – either organic or artificial – is the driving mechanism for progress, and both are increasingly "decoupling from consciousness."

From a more concrete perspective, recent ground-breaking research by James Tee and Desmond Taylor confirms that information-processing procedures within the human brain are digital. Here, information is processed in a discrete, rather than a continuous (analog), form.[9] This infers that it may indeed be possible for computer-generated artificial intelligence to seamlessly coexist with the human brain. The possibility that we possess digitally-structured brains gives hope to the prospect of intelligence augmentation, where machine intelligence and human intelligence can complement each other. Instead of designing better stand-alone machines, intelligence augmentation focuses on designing devices that help humans think better. We win the game by putting the new 'marquee' player inside our heads! A technologically inspired revolution of being through brain-computer interfaces. As the interbreeding of Homo sapiens with Neanderthals and Denisovans created the first great leap forward in human endeavour, the second 'great leap' will result in a cognitive revolution inspired by the machine.

Figure 11: Evolutionary biologist Sir Julian Huxley is regarded as the father of transhumanism. In 1957, he boldly declared: "The human species can, if it wishes, transcend itself—not just sporadically, an individual here in one way, an individual there in another way, but in its entirety, as humanity."

A Human(e) Alternative

Up until recently, artificial intelligence, like natural selection, has evolved through a 'bottom-up' process where competence is gained without comprehension. Twentieth-century computer systems turned massive amounts of data into comprehensive bits of information from which humans would make informed decisions. Machines did the job without understanding or informed purpose. In contrast, human reasoning and subsequent comprehension of information – although itself a product of blind natural selection – can create top-down 'intelligently designed' adaptions.

The elusive dream with respect to AI development is that one day we will invent machines capable of top-down complex reasoning. This may only be possible, Daniel Dennett contends, through a symbiotic collaboration between cultural evolution which can enable "more effective design processes, foresighted and purposeful and dependent on the comprehension of agents: intelligent designers." That said, Dennett concedes that advancements in AI so far have yielded impressive results through bottom-up processing, akin to the unknowing amoral powers of natural selection: "Machine learning systems that use bottom-up processes to demonstrate once again the truth of Orgel's Second Rule: Evolution is cleverer than you are." [10]

Both Daniel Dennett and Richard Dawkins take a softer approach than Harari's somewhat dystopian vision of 'organic' humanity. Both may be called evolutionary compatibilists; that is, they believe that although our genetic heritage greatly determines how we behave, we also possess the ability to act freely against our naturally evolved impulses. For Dawkins, it is our newly-evolved capacity for super-niceness that enables us to rebel against our selfish nature. For Dennett, it is our culturally-evolved ability to rationally consider different options – either consciously or unconsciously – that appear from predetermined impulses generated in the brain.[11] In other words, both men are saying that yes, our desires to behave in certain ways are the predetermined result of our genes; but we now have the ability to freely adjust or even reject the urge to act impulsively.

These culturally-evolved adaptations, though they originate from the ordered chaos of natural selection, enable us to escape the confines of our genetically-inspired urges. The validity of this assumption is important. If true (which I believe it is), then metaphysical notions of duality between mind and body, and religious teachings about good and evil, are not only redundant but absurd. Two and a half thousand years of philosophical musings and religious dogma are thrown out the window. Instead, we are prisoners of our inherited genetic past, programmed to survive and reproduce; but we can, to some degree, transcend our predetermined nature with our newly culturally-evolved ability to make informed moral decisions. It is therefore no surprise that Dennett and Dawkins pay tribute to the Western tradition of liberalism and the cultural memes put in place to ensure a free and ordered society which is accepting of secular values.

Our individual brains may be finite, organic creations, but our collective knowledge and cultural adaptations can create what Dennett terms "distributed comprehension,"[12] where previously unsolvable mysteries can now surrender their secrets to the logic of science. Add to this our evolved capacity for our conscious minds to imagine new worlds, separate and to some degree free of our instinctual impulses, and we may reach for something more than this: where previously, there were seemingly unsolvable (even mystical) mysteries, there is now a new frontier of being, based firmly on material ground. As Dennett concludes:

> where we currently are, does seem to be a major game-changer, permitting us to create, by foresighted design, opportunities and, ultimately, enterprises and artifacts that could not otherwise arise. A conscious human mind is not a miracle,

not a violation of the principles of natural selection, but a novel extension of them.[13]

Any notion of Cartesian dualism – the dual domains of mind and matter – is an illusion. Human consciousness comes from representational activities within the brain that speculate appropriate reactions to those activities. In other words, thinking about what has happened and what we can do about it (our sense of consciousness) are derived from Darwinian evolution, not some magical gift from the Gods.

If any understanding can be gained from the exponential increase in knowledge about how the natural world works, and from our increasing ability to transcend the confines of natural selection, we must first objectively observe these newfound discoveries and capabilities, and only then make informed moral decisions. It may be no mistake that the discovery of DNA and the subsequent reality of gene sequencing coincided with new 'emergent' directions in human values and behaviour. These events have also coincided with a crisis in how people practise their desire to live a spiritual life.

If we are bluntly honest with ourselves, notions of a personal God who demands prayer and sacrifice is absurd. That said, even today, where the truths of scientific enquiry have all but decimated old myths regarding the natural world and the disasters that inevitably occur, people will still bring forth the talismans of hope, and invoke a desperate inner plea that things can somehow be different. For many people, the idea that our actions and aspirations are not guided or judged by God is a truly frightening proposition. God did not inspire you; nor did the Devil torment you. Supernatural notions of good, evil, and divine intervention are an illusion; instead, we are all born burdened by our genes. Our own uniqueness originates from brain activity; self–awareness and consciousness live exclusively inside our brains.

Taken one step further, any notions of a divine and immortal soul are also tenuous. Traditionally, the soul represents humanity's search for an inner life independent of the trappings of material existence. Religions and even the concept of God(s) have been created to act as a bridge of understanding (and as a control mechanism) in creating a meaningful inner life. However, the idea that we are spiritual beings has outlived the idea of God, and many people are now concerned with the search for inner meaning outside the trappings of religious faith.

The term artificial intelligence was first used by John McCarthy in 1955. Today, intelligent machines (including computers) consume over ten percent of all electrical power. Our superior intelligence has not only created machines that can support our technological aspirations; new forms of artificial intelligence may indeed surpass us and form concentrations of 'super intelligence.' The first evidence for this is in front of our eyes. As Steve Wozniak objectively proclaimed:

> All of a sudden, we've lost a lot of control. We can't turn off the internet; we can't turn off our smart phones; we can't turn off our computers. You used to ask a smart person a question. Now who do you ask? It starts with go…. and it's not God.[14]

The evidence for superior intelligence in human endeavour is obvious – only a community of highly-intelligent organisms could travel to the moon and return safely back home. The evidence for artificial super intelligence is emerging, and the seeds of it shine seductively and directly at us. Whether we are hypnotised into submission, or enlightened to become more than we ever dreamed, is up to us – personally and communally. And everything may depend on seeing ourselves as empowered beings fuelled by a sense of imagination, spirituality, and a desire to do 'good.'

The candle of faith, and adherence to the laws of religions composed thousands of years ago, will not guide us in a world illuminated by the light of our machines. Our religions and most of our philosophies have failed us. But we cannot deny that, for most of us, there remains a yearning for what Christopher Hitchens referred to as 'the numinous' – life events that arouse spiritual emotions that can be awe-inspiring. This yearning desire appears to be hardwired in all of us, and it is not surprising that religions, totalitarian regimes, and even New Age hucksters have used it for their own ends. Maybe it is time to also recognise this uncomfortable truth. But it may also be time to recognise that our spiritual struggles are not totally in vain.

Key Points

- Superior intelligence was once exclusively a human trait; this is no longer the case. Artificial intelligence is fast reaching the point where it will independently surpass that of humanity in all areas of intellectual endeavour (the singularity).

- Artificial super intelligence will have the ability to solve problems previously incomprehensible or unsolvable by humans. Organic human reasoning may turn out to be not only inferior, but also redundant. This means that any future upgrading of the human form will also most probably involve some form of AI.

- It is important that we develop the emotional and technical capacity to create forms of AI that are compatible with our material and moral desires, where AI and humanity embrace each other as fellow sentient beings. The first step on this journey will be the rejection of supernaturalism, and the abandonment of any idea that individual human consciousness exists independently of the material domain.

- As a species, we are at the cusp of a second great leap forward that will result in a cognitive revolution inspired by the machine. Belief in religious supernaturalism is not only absurd in this paradigm, it is also destructively delusional.

- Our emerging 'super intelligence' is not supernatural and may indeed represent extraordinary evidence that humanity – or some augmented form of it – will transcend materiality.

- This vision will be dependent on our ability to create a symbiotic collaboration between cultural evolution, our spiritual desires, and the intelligent designers of this brave new world.

- What is transpiring now is not only the possibility to reach for things known, but to also take steps into an unknown future, towards worlds of being, that were once thought to exist exclusively in the lap of the Gods.

PART B: FROM RELIGION TO SCIENTIFIC SPIRITUALITY

Think of the distressing contrast between the radiant intelligence of a healthy child and the intellectual feebleness of the average adult. Is it not at least possible that in fact religious education is largely to blame for this relative atrophy?...No our science is not an illusion. What would be an illusion would be to think we might obtain elsewhere that which science cannot give us.

Sigmund Freud, *The future of an Illusion.*

FIVE

A Spiritual Species

Do not undertake a scientific career in quest for fame or money. There are easier and better ways to reach them. Undertake it only if nothing else will satisfy you; for nothing else is probably what you will receive. Your reward will be the widening of the horizon as you climb. And if you achieve that reward you will ask no other.

Cecilia Helena Payne

Introduction

If the validity of scientifically-generated truths is so compelling, why do many of us still cling to the promise that a divine force has a special interest in who we are and what we become? Recent research indicates that, to some degree, each of us is genetically tuned to yearn for something beyond tangible materiality. We all have an instinctual desire to find meaning and purpose – in essence, to be spiritual beings. Our spiritual journey, like our moral arc, has been chaotic, violent, and continually usurped by jealous and powerful self-interests. This instinctual yearning has also been culturally manipulated and transformed into a communal control mechanism, where ruling elites have held sway over what we believe and how we live our lives.

I will attempt here to first investigate if our spiritual yearning is indeed a heritable trait, and will show that this trait, like other traits, is embedded in our ancient past. Whether interpreted through the lens of philosophy, the diminishing blind faith of religious belief, or alone, all of us knowingly contribute – either productively or destructively – to the cultural evolution of our species. We are the only living organism to do this. Our persistent yearning to be spiritual and to search for more than this, is not only programmed into our genetic nature but is also present in every cultural construction through our expressions in art, religious worship, architecture, literature, and more abstract forms of spirituality. Unfortunately, this powerful and creative trait has also been culturally warped – even in very ancient groupings of people – to dominate, rather than illuminate.

Is Religion in our Genes?

In 2004, geneticist Dean Hamer authored *The God Gene: How Faith is Hardwired into Our Genes*. The book detailed a specific link between a particular gene – the *VMAT2* gene – and human behaviour. Hamer further contended that a person's predisposition towards spirituality is also influenced by genetic factors.[1] His controversial claims were mostly either ignored or ridiculed by leaders of religious subcultures. It became immediately clear that any assertion of a genetic link to personal spirituality and religious belief is both contentious to many people and confronting to those who dare to even think about it. How we view our inner self and our role in the world has traditionally hinged on which religion or spiritual compass we have freely chosen. Christianity, Judaism, and Islam even demand that their followers 'live' their faith every day.

Hamer's work was confirmed by later research led by Dr Laura Koenig in 2008, which suggested that the heritability of religiosity (the desire to outwardly express spiritual beliefs) becomes more apparent as a person ages and casts off earlier environmental influences.[2] Additional, ground-breaking research by behavioural geneticists Gary Lewis and Timothy Bates, in 2012, confirmed again that there is indeed a genetic component to our desire to connect to a spiritual force, either collectively or on our own.[3] A 2020 meta-analysis of nature and nurture in religiousness across the human lifespan confirmed beyond reasonable doubt that there is a heritable component for spirituality.[4] This discovery now brings into focus the critical question of how the trait towards religiosity has evolved both naturally and through changes in cultural influences over time. Can we make any conclusions about this unique urge that separates us from every other organism?

Harmer also drew from previous work of socio-biologist Edward O. Wilson, who argued that 'group selection' (the complicated intermixing of our genetic tendency towards spirituality as it is filtered through a communally-accepted set of sacred practises/rituals called religion), rather than individual or kin selection, is primarily responsible for an individual's religiosity. Wilson put forward the idea that the way a person conforms to the group dynamics of religion is in itself adaptive. Controversially, group selection infers that the unit of selection is the group, rather than the individual. Here, a group of individuals working together achieve a higher degree of 'selection fitness' than if the individuals were all alone.

However, any inclination for the idea that group selection was part of Darwinian natural selection is aggressively opposed by Richard Dawkins (and a majority of other evolutionary biologists) who hold firm to the pre-eminence of individual selection. For Dawkins, there are "two units of selection": vehicles (the living individuals that survive long enough to pass on genetic information) and replicators (the actual genes that have been delivered by the vehicle). Irrespective of each viewpoint, it is obvious that for many of us, religiosity is part of who we are and how we live. Interestingly, Dawkins in 2020 released the following quote by Wilson on Twitter: "The real problem of humanity is the following: we have Palaeolithic emotions, medieval institutions and God-like technology."

If we accept that spirituality is a heritable instinct, then this trait is extremely personal and most probably comes from our desire to withstand the inevitability of death that surrounds all living things. Our communal desire to be religious, in contrast, is a culturally-created meme which evolves in accordance with the power structures within a society at a particular point in time. It is therefore possible (even likely) that a person can be spiritual without the need to identify with any structured religion. For Dean Harmer, spirituality is defined and measured by our level of self-transcendence: that is, by our ability to observe the nature of the universe – and our place in it – outside the confines of our own personal needs and desires. In other words, to reach out beyond ourselves. Although the trait of spirituality is genetically sourced, the instinct is variable between people. Interestingly, a 2017 study found that deliberate suppression of a person's tendency towards spirituality increased their risk of depression.[5]

Probably the most significant human disposition that arises from the uneasy blending of genes and culture is the urge for us to make sense of where we come from, why we are here, and what happens when we die – in others words, we are spiritual by nature, and religious through nurture. Religion also serves as a conduit whereby individuals are persuaded to subordinate their immediate self-interest to the interests of the group in the name of their God. However, in most cases, it appears that any altruism inspired by religious beliefs does not apply with respect to tolerance of other religions or cultures. A survey of religious beliefs and practises from a distinctively Western perspective reveals that while we are definitely spiritual by nature, our religious dispositions have evolved according to cultural determinants which are inherently violent and mostly patriarchal.

From an historical perspective, religious beliefs and practices would become the centrepiece for functioning ancient communities. Religious practice was usually experienced in a communal setting with agreed rituals and a unified understanding of what was sacred to the community. Renowned French sociologist Émile Durkheim said that, "A religion is a unified system of beliefs and practices relative to sacred things."[6] Central to the idea of 'sacred things' is the belief in a divine God or Gods who interact with humanity within a unified natural world that includes both divine and human realities. Therefore, the practice of religion involves a deliberate and organised attempt at finding meaning in the experience of our lives. It acts both as a guide and a signpost to the possibility of linking with a creative essence known as the divine or, more directly, a God. Any God figure is a supernatural force that can focus religious experience between an abstract, divine realm and the human realm in which people live their everyday lives.

Throughout history, religion has been a tribal binding mechanism to the extent that it becomes either the sword of the victor or the shield of the defeated. However, even given our communal religious dispositions, history tells us that, at best, we are only a moderately social species outside of kin and tribal alliances. From what we know about our ancient past, it is apparent that we were only social to the extent that being social allowed us to pass on our own genes and those of our kin. We cooperated because it was in our own self-interest to do so. If this premise is taken as true, then any moral codes of conduct (including supposedly God-given laws) have originated from the drive to replicate favoured genes across time and cultures. The religions created to service or manipulate our natural desire to be spiritual are man-made institutions which have culturally-evolved over time.

A Primal Desire

The idea and hope that some type of supreme and guiding force surrounds us has permeated humanity since the earliest expressions of civilised thought. For want of a better term, I have settled with the word 'God' when referring to this divine ultimate construction. Scholars and philosophers have attempted to understand God in various ways. In its most powerful incarnation, God is said to be both immanent in that he (given the masculine image of God in most Western cultures) permeates everything within the natural universe. He is transcendent – he can rise above and is independent of worldly confines. God is also all-powerful (omnipotent) and all-knowing (omniscient). Finally, this divine being is outside of the material concepts of space and time, and therefore beyond the comprehension of mere mortals. It is usually this last 'God condition' that is used by some God apologists to negate any attempts at rational discussion about his existence.

The first writings and symbols of ancient civilisations were, in most cases, religious in expression and context. These expressions illustrated the need for a belief system that extended beyond ordinary human experience. Therefore, any religion or religious belief needs to introduce the premise of a divine realm that has within it a creator who transcends human existence and the laws of nature. For the religious, there has to be 'more than this,' but only under the all-seeing eye of God. The first representation of God could be seen in the sun's daily birth (sunrise) and death (sunset). People were aware that it was the sun that gave light, warmth, protection, and new life through spring and summer. It is not surprising that the daily warmth and light of the sun were venerated as sacred and miraculous gifts. The sheer terror that darkness would bring at the end of each day could only end with the dawn of the resurrected sun. God was inexplicably linked to the workings of the natural world. Where there are no defined and understandable laws of nature, the deeds of God take pride of place in human life and death. Even today, a belief in an all-powerful, loving, and personal God acts as a shield against those things beyond our control.

The first tangible evidence of some kind of religious belief in the Western tradition comes from the Old Stone Age, the Palaeolithic era; early proto-human Neanderthal clans carefully buried their dead around 100,000 years ago. The deceased would be given tokens of faith and their bones would be preserved for some ceremonial purpose. Neanderthal clans may have also gained an elementary grasp of the existence of a spirit world, where the dead person's life force could be given back to the living, as evidenced by such acts of ceremonial cannibalism as privileged members of the clan eating the deceased's brain.

Towards the end of the Old Stone Age, around 30,000 to 17,000 years ago, early humans began to express their spiritual beliefs through cave paintings and elementary artefacts, such as clay statues that emphasised nature and fertility. The dead were given more sophisticated burials, as evidenced by a deliberate ceremonial placement of the body. These ceremonies included group burials and the reburial of the deceased's bones. Interestingly, this practise probably emerged from the inter-mating of Homo sapiens with Neanderthal and Denisovan hominoids around 50,000 years ago.[7] Our brains became bigger and envisaged something more than mere survival. By the end of the Old Stone Age, religious art began to reflect worship of a mother Goddess figure, who represented fertility and the changing seasons of life. Our desire to understand the consequences of death meant that religiosity had culturally-evolved to focus on some form of afterlife. Interestingly, recent research indicates that genetic dispositions for aggression and hyperactivity appear only in Homo sapiens and not within the known genome of either Neanderthals or Denisovans.[8] This may be one of the reasons that these hominids became extinct – they were hunted down by our more aggressive ancestors.

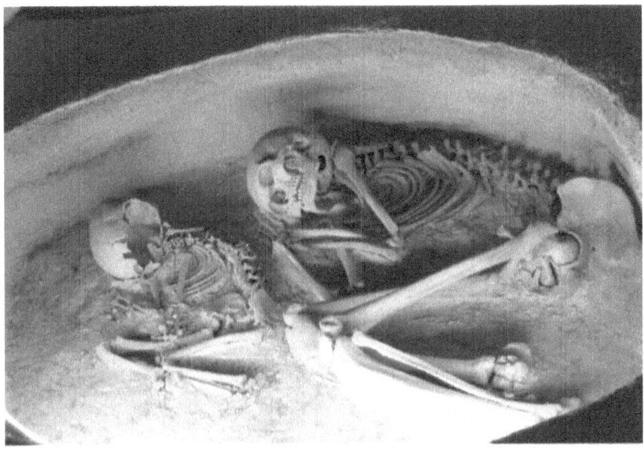

Figure 12: Late Palaeolithic era remains show ritualistic burial of the dead.

The New Stone Age or Neolithic era (9000–3000 BCE) saw the transition of hunter-gatherer societies into small villages that were dependant on domesticated crops and animals for survival. Humans had finally gained some limited control over the natural world. However, this small degree of influence remained dependent upon the changing seasons and the cycles of the natural environment. The concept of fertility and the essential life-giving relationship between nature and the cycles of birth and death increased the spiritual power of the mother Goddess figure. Religious rituals were interwoven with the seasons and with the passage of human life through birth, adulthood, and death. Neolithic culture may have also identified the act of sex as a sacred ritual.

It could be argued that the twin occurrences of the emergence of large, civilised groups of people and the beginnings of more complex religious beliefs were not coincidental but rather interdependent. Large groups needed to communicate effectively both on an individual basis and communally. For the first time, human beings could communicate in abstract terms that set them apart from all other animals. They now had the ability to also communicate information about things that may not exist. The emergence of legends and myths gave humans a template for living in harmony with each other and with the natural world. The exact cause of this sudden evolution in cognitive thinking is unknown but, as previously mentioned, it is also most likely related to the intermixing of genes, an improved diet, and the subsequent increase in the size of the human brain. This 'great leap forward' happened virtually at the same time in communities that had little contact with each other. A creative spark of civilisation and religious desire had taken hold of humanity. Our spiritual nature had now culturally-evolved to accept the interdependence of a supernatural domain and the material world.

Figure 13: Early figurine of a mother Goddess figure from the region now known as Western Turkey, dating back to around 5750 BCE.

A Man's World

Central to any conception of a civilised society is the premise that humans will cooperate with each other — either voluntarily or by coercion. Daniel Dennett believes that "human cooperation is a delicate and remarkable phenomenon, and indeed quite unprecedented in the natural world, a unique feature with a unique ancestry in evolution."[9] Voluntary cooperation is dependent on mutual trust that extends outside of evolutionary instinctual boundaries. For Dennett, the driving force behind the cumulative material advancement of humanity is our ability to communicate through language. "Language, as we have seen, is the key invention, and it expands our individual cognitive powers by providing a medium for uniting them with all the cognitive powers of every clever human being who has ever thought."[10] However, it must be conceded that the road to some form of ordered civilisation has been slow, violent, and dominated by the male form of our species.

As communities became more complex, the worship and relevance of the mother Goddess figure diminished. A disturbing aspect in most of the religious creation myths of antiquity is the amount of violence – in particular, patriarchal dominance and family destruction – that the Gods and early humans inflict upon each other. (Appendix 1 at the end of this book details the main creation myths of early Western civilisations.) In Egypt, Seth kills his brother Osiris. The great God of Mesopotamia, Marduk, disposes of his consort, Tiamat, by splitting her in two to create the earthly and divine realms. Cain murders Abel in the Old Testament. In the Greek creation myth, Zeus kills his father, who has previously devoured all his other children – without doubt the most violent of these stories. Finally, Romulus eradicates his brother Remus in order to be the sole ruler of Rome. These violent acts are usually between father and son or between rival siblings and, as such, may contain a deeper meaning as a representation of complex social and cultural themes that have permeated the history of humanity. Sigmund Freud contended that ancient myths involving the murder of a father by a son or a brother by a brother reflect an instinctive (genetically-conditioned) prehistoric struggle for dominance. Freud argues that within the context of a small, family-based tribal unit:

> The strong male was the master and
> father of the whole horde: unlimited in
> his power, which he used brutally. All

females were his property, the wives and daughters in his own horde as well as perhaps also those robbed from other hordes. The fate of the sons was a hard one; if they excited the father's jealousy they were killed or castrated or driven out.[11]

In time, however, the youngest son would usually get the opportunity to usurp his aging father. This son, usually the favourite child of the mother, will become the father and leader of the tribe. Some very early rituals indicate cannibalism of the father's body as a means of transferring his spirit to the tribe. The old father was not only feared and hated by his tribe, but he was also honoured for his position and power.[12] If we can assume that these religious myths are grounded in prehistoric truth, this adds depth and complexity of meaning to the ancient stories on how things came to be. Sigmund Freud was a man of the early 20th century and, as such, was unaware of the knowledge given to us by genetic science and evolutionary biology. He did, however, conclude correctly that many of our desires are grounded in ancient instincts. Our selfish genes may indeed be very selfish if left to their own amoral desires. Although tribal and kin loyalty existed in the ancient world, an 'alpha' male always held sway when it came to replicating his genes; and this would be reflected in the patriarchal control of religion.

It is not surprising that the first practice of religious/magical rituals coincided with the development of early tribal communities. These rituals were not only attempting to commune with the divine, but they may have also originally represented a joining of individuals for the good of all. Religious practices may have also been the first examples of the emergence for ultra-truism which transcended genetic obligations to one's kin. However, as previously discussed, allegiance to one's tribe meant that other tribes were automatically excluded and targeted. Religious rituals could have been used to form a style of tribal 'group think' that was ultimately manipulated by clan leaders driven by selfish motives and selfish genes. Looked at from this perspective, religion may have initially been an expression of hope for something more than selfish survival. It gave a promise that things may change, and our violent nature might be subdued. The reality of this is unfortunately the opposite; religious beliefs in a supernatural deity and the blind obedience given to a God-chosen representative – usually a dominant male – made religion a culturally-inspired weapon for the replication of those selfish genes best suited to ruthlessly rule over others.

For over 3,000 years, the oldest Western civilisation – that of the ancient Egyptians – used the motif of an all-powerful King (Pharaoh) as the divine representation of the Gods on Earth. Images of Pharaoh smiting his opponents and dominating his subjects are first seen in the Narmer Palette (dated to around 3100 BCE). This disturbing image was transposed in the likeness of many reining Pharaohs and endured all the way through to the Alexandrian (Greek) dynasties, just before the birth of Christ. On the rare occasions where a woman ruled as Pharaoh, she is usually depicted wearing the masculine beard of kingship. As the religious leader of his people, Pharaoh had one job: ensuring that the sun was 'resurrected' each day. In return for performing this daily miracle, Pharaoh was worshipped as a living God. Even in death, he still performed this task. If there is indeed a genetically-coded desire for spirituality, it has been commandeered by those people (usually men) in power, and used to rule, rather than serve. It is also not surprising that those in power ensured the royal bloodline of the king remained genetically pure, such that incest was the preferred method of royal reproduction. Everything and everyone had to serve the divine, royal bloodline. The human 'Gods' of the ancient world were not all-knowing, or all-loving; they were not even all-powerful. But they were hell-bent on ensuring the immortality of their genes at any cost!

It is now clearly apparent that what ancient cultures thought were kingly acts of divine order or retribution were really the result of genetic influences – through the omnipotence of natural selection – and a man-made cultural environment focused on maintaining elite blood lines through patriarchal power and dominance. It may also at least partly explain the inherent violence that is still with us in most Western cultures today, where power, money, and the material trappings of society are valued above all other things. It may mean that, as a culture, we need to rethink how we gain meaning from our material existence. As John Avise concludes:

> the suggestion that there exists a genetic predisposition for human belief systems does not prove that religion has not caused tremendous human suffering. If belief systems have profited our ancestors evolutionarily, in part it has been through the moral justification of violence against dissimilar cultures.[13]

Most cultures, ancient and modern, have also recognised (and acted on) "the susceptibility of children to religious indoctrination. A child's concept of life and death are vague and at times contradictory. The human mind seeks to rectify contradictions, and religious education may serve that function in young children, if in an irrational fashion."[14] In other words, we are born with an innate spiritual yearning that can be quickly manipulated and corrupted by religious doctrine and dogma. In most cases, this delusion lasts a lifetime. Nature (our genes) loads the gun, and nurture (our cultural context) pulls the trigger. That said, our yearning to engage in expressions of spirituality is a major factor that separates us from other organisms. It has also provided the means by which we can envision new frontiers of human endeavour. Over time, people have either redirected their spiritual urges or surrendered them to the prison of structured religious belief. We cannot deny this heritable trait, nor the ability of cultural institutions to manipulate it for their own means. If we allow religious organisations into our schools, then equal time should be given to encouraging personal empowerment of each young person's desire to be spiritual without any dogmatic playbook of faith.

Figure 14: The Narmer Palette, dating to around 3100 BCE, shows the first King of a united Egypt smiting his enemies.

Figure 15: Much the same image, but this time dating to the 'Greek' Pharaohs nearly 3,000 years later. This period of time spans over half of known civilisation on Earth. Pharaoh was King, God, warrior, and the embodiment of everything divine in Egyptian religion.

Key Points

- Each of us is genetically tuned to yearn for something beyond tangible materiality. We are by nature spiritual beings.

- A person can be spiritual without having the need to identify with any structured religion.

- All forms of structured religion are man-made, culturally-created memes which have evolved in accordance with the power structures within a society at a particular point in time.

- Spirituality, however, is defined and measured by a person's level of self-transcendence, that is, by our ability to observe the nature of the universe – and our place in it – outside the confines of our own personal needs and desires. In other words, it is defined by our ability to reach out beyond ourselves.

- We are born hard-wired to yearn and search for something more than this.

- Although our religions may have initially been an expression of hope for something more than selfish survival, they also nurtured a deluded meme that a personal God ruled over everyone and everything.

- Religious beliefs in a supernatural deity and the blind obedience given to a God-chosen representative – usually a dominant male – made religion a culturally-inspired weapon for the replication of those selfish genes best suited to ruthlessly rule over others.

- Although this trait has been culturally warped and abused, it still pushes each of us to search for meaning.

SIX

A Rational World

There are indeed two attitudes that might be adopted towards the unknown. One is to accept the pronouncements of people who say they know, on the basis of books, mysteries or other sources of inspiration. The other way is to go out and look for oneself, and this is the way of science and philosophy.

Bertrand Russell, *Wisdom of the West.*

Introduction

One of the few good things that resulted from the ritualistic practises of organised religion in the ancient world was that people could observe if the supernatural spells cast on their behalf actually worked. Somewhat paradoxically, these magical acts encouraged empirical observation, which is still at the heart of science. The smarter priests and shamans would also observe the world around them and use the information gained to accurately predict natural events. As the chosen intermediary between mortals and the Gods, a priest could justify his power and influence by 'magically' predicting what were really natural occurrences.

For example, the later Neolithic inhabitants of Britain, Western Europe, and parts of Asia Minor began to construct stone megaliths. The oldest of these can be found at Göbekli Tepe, located in modern-day Turkey. These large stone monuments may have been arranged to reflect and measure the changing seasons and the movements of the sun. Circular megaliths, such as Stonehenge, mark the crucial times of the summer and winter solstices. Stone megaliths also served as ceremonial, religious, and burial sites, where rituals could be stage-managed by a priest at critical times of the year. This recognition of the earth as a dynamic and ever-changing environment meant that New Stone Age society attempted to understand and influence the cycle of nature. Therefore, early religious and spiritual practice centred on the gaining of spiritual power—*mana*. *Mana* could be gained through ritualistic practices that celebrated fertility and movements in the natural world, such as the changing seasons and the positioning of the sun, moon, and stars.

It would, therefore, be reasonable to assume that some people also sought to observe the motions of the natural world outside the confines of religious ritual. From a Western perspective, the idea that there could be rational (natural) explanations for how things came to be originated during the seventh century BCE in ancient Greece. A few wise thinkers rejected magical (supernatural) ways of understanding the world and sought to find methods of investigation which relied on observation and objective reasoning. They did not abandon their Gods or their religions, but they at least considered the idea that many of the workings of the natural world could be understood as elements in reasoned order. This radical departure would greatly influence the naturalistic philosophy of later classical Greek and Roman thought.

However, it would not be until the great Renaissance that a formal scientific way of observing and investigating how things are would be reimagined by thinkers in search of rational ways of understanding the material world. Francis Bacon would give us his *Novum Organum* – a 'new method' of epistemological authority based on the authority of experimental science.[1] Most importantly, this time marked a return to the separation of religious beliefs from the workings of science. It also marked the beginning of (and in some ways a return to) the quest for personally inspired spiritual enlightenment outside the confines of Church and state. For the first time in history, God-given laws gave way to "the light of experience and the discipline of experiment" as the primary arbiter of truth and knowledge in the natural world.[2]

The Birth of Reason

The ancient Greeks were the first civilisation to look inwards at themselves to find meaning and understanding. Humanity, through the use of reasoned thought and action, could find a pathway to comprehending the material world. The implementation of democratic freedoms also encouraged different ways of viewing what constituted a good life. Their pantheon of Gods was now supported by a collection of reasoned thoughts. The quest for knowledge resulted in competing and even binary philosophies. However, underlying this tapestry of varied philosophies was the desire for moral integrity that was essentially gained through the experience of living a productive and harmonious life that supported the broader ideas of the city-state.

For the first time in Western civilisation, simple worship of the Gods with prayer and sacrifice was not enough. Rather, the ancient Greeks desired to understand the forces that permeated their lives. To gain understanding, humanity had to question and participate in the search for truth and meaning. This active participation was the first step to creating a scientific understanding of cosmology – where we came from. Our culturally-evolved capacity to rebel against our selfish nature was also directed inwards, to search for what constituted a meaningful life. Morality was more than merely serving the Gods or his designated intermediaries. Unlike the earlier civilisations of Egypt and Mesopotamia, many educated Greeks sought rational and spiritual illumination. For the first time, science and spirituality were part of a greater narrative that became known as philosophy.

From the perspective of the ancients Greeks, philosophy was simply the love of wisdom, and a philosopher was someone who actively pursued truths – both relative and absolute – about the natural world. The first of these were known broadly as the pre-Socratics. They were the first people to formally abandon purely supernaturalistic explanations and search for 'the essence of things.' Among the pre-Socratics were the first professional wise men, the Sophists, who declared that "Man is the measure of all things."[3] It was humanity that provided both the compass and the template for living a good life. The Sophists, led by Protagoras, pioneered what became known as a subjectivist way of viewing the world. In this view, there is no absolute truth; rather, truth is dependent upon how each individual perceives the world from his or her own unique perspective.[4] Therefore, there were no absolute truths in religion or moral codes of conduct.

By the sixth century BCE, Greek philosophers also sought to find meaning through naturalistic explanations. Anaxagoras observed that the sun and moon are natural objects, rather than celestial Gods. Protagoras even honestly admitted that divine entities might not exist. But change did not come easily; both men were banished and died in exile. It became clearly apparent that to challenge religiously-inspired ways of understanding things was a very dangerous undertaking. However, some individuals held strong to the light of reasoned thought.

Pythagoras, probably the most significant of the pre-Socratic philosophers, stands in contrast to the naturalistic approach that would dominate later Greek philosophy. Instead, he viewed the universe from an abstract, even transcendent perspective. For Pythagoras, it is the realm of mathematical abstractions that describe material reality perfectly. Best known for the formula $a^2 + b^2 = c^2$, he correctly argued that no material rectilinear triangle could match the mathematical perfection of the equation. From this, he concluded that there must be a creative force which has designed everything prior to its material representation. Therefore, mathematical abstractions predate physical discoveries. This way of seeing our universe also predates theories of cosmological natural selection with intelligence by over 2,600 years, but infers much the same vision of how things came to be the way they are: our universe is comprehensible to us because intelligent beings (similar to us) designed it.

Pythagorean ideals would later greatly influence Plato and his vision of the cosmos. The followers of Pythagoras placed supreme reverence on the "divinity of number." The numbers 1,

2, 3, and 4, known as the tetraktys, not only added up to the 'sacred 10' but also represented the point, the line, the plane, and the solid, respectively. Interestingly, Pythagoras did not abandon supernaturalism. Like Plato, he even believed in the transmigration of souls (reincarnation). Pythagoras believed that a separate and divine soul existed within each human individual. This divine spark, or *psyche*, was trapped and could only be released by a process of purification and denial of bodily desire. Release could take several lifetimes to complete.[5]

Opposed to this transcendent world view was the emerging philosophy of Socrates. He believed that absolute truths did indeed exist and that the challenge for humanity was to discover these truths. For Socrates, truth could only be found through questioning and challenging ourselves and others on all aspects of intellectual discourse. A universal truth lay hidden in men's minds; the real adventure of life was to reveal this through a process of challenge and response. This process has become known as the Socratic method.[6] Unfortunately, Socrates was accused of corrupting the youth of Athens and was sentenced to death by poisoning. On the subject of death, he said, "Wherefore be of good cheer about death, and know of a certainty that no evil can befall a good man, either in life or after death."[7] Socrates died in 399 BCE as a martyr for truth and adherence to the laws of Athens.

Socrates's teachings directly influenced the emerging philosophies of the Cyrenaics and the Cynics, albeit in different ways. The Cyrenaics were a pleasure-seeking cult who believed the true aim of life was the attainment of pleasure within the constraints of traditional Greek moderation. Real truth could be found in the simple enjoyment of life. The Cynics, however, interpreted Socrates's teachings as saying that real truth can only found through the abandonment of the material trappings of civilised life. The main legacy of Socrates's teaching comes to us in the philosophy of his greatest student, Plato. Plato was a student and follower of Socrates for the final eight years of Socrates's life. Socrates's emphasis on the importance of the Polis as a civilizing force should not be understated. He was the first person in Western civilisation to recognise that the culture created to nurture communal living was the only thing that restrained the chaotic nature of individual citizens. For Socrates, the validity of state-enforced laws provided the rational power by which to minimise individual corruptibility.

Plato was from an aristocratic Athenian family who were direct descendants of the great law-giver Solon. After a failed attempt at creating a utopian society, Plato returned to Athens to open his academy. During this period, Plato wrote around twenty-four distinct dialogues that captured the essence of his philosophy. Most modern interpretations of Plato's works are from a holistic perspective, so a broader sense of meaning can be gained by reading the dialogues. That is, it is best to understand the combined themes of his teachings. Esteemed academic Georgia Harkness best describes how to understand Plato's contribution. She says, "Many of his most important ideas are couched in myth, as great religious truth is apt to be. After one reads for a time, one begins to grasp intuitively Plato's major emphasis rather than found in specific sentences."[8]

Harkness has no hesitation in linking Plato's philosophy directly to spiritual prose. Plato guides his followers and provides a template for leading a meaningful and good life. Harkness contends, with regards to the fundamental principles of Plato's teaching, that:

> Among them are that the basics of all existence is a moral order: that man's chief task is to lead the good, which is also the beautiful, life; that this is to be measured by health and harmony of soul; that the state exists for moral ends; that no morality, whether individual or social, can be attained apart from reason.[9]

Plato also refined Socrates's concept of universal truth. Plato contended that absolute truth can only be found in the realm of reason. The material world is a false reality, where imperfect reflections of the truth provide only a distorted view of how things really are in the universe. This idea is probably best expressed in Plato's famous allegory of the cave. In this story, prisoners are trapped within a cave and can only see reflections of the world around them. They will never see the world in its true light. As Harkness concludes:

> The truly real is neither the material world nor a world of human dreams. It is an eternal and ideal goodness, truth and beauty in which man, as he lives at his best, may participate, but which he may never fully fathom or encompass.[10]

In his vision of what the perfect city-state would look like, Plato proposes that only those citizens chosen as superior are fit to breed. Family units were to be discarded and all children were to be, in essence, the property of the state. This very early expression of selective breeding of humans is undoubtedly the worst use of planned eugenics, akin to the Nazi programs. From Plato's *Republic*: "and if we are to have a pedigree herd, mate the best of men with the best of our women as often as possible, and the inferior men with the inferior women as seldom as possible."[11] These words were written nearly two and a half thousand years ago in a book that is widely acknowledged as the first text on political economy. I do not intend to diminish the importance of Plato's contribution to philosophical thought, but it must be acknowledged that ideas about eugenics are not new and represent yet another uncomfortable truth. As previously mentioned, Plato also did not abandon his religious convictions, but he did inspire us to search for the light of higher truths through reasoned thought.

The third great philosopher of the Classical age was Aristotle, a pupil at Plato's academy. In 343 BCE, Aristotle was summoned by King Philip of Macedon to tutor the then-13-year-old Alexander. He later returned to Athens to open his own academy. Unlike Plato, Aristotle had a greater interest in the material world and in the 'scientific' truths that could be gained from empirical study of the natural environment. While Plato could be termed an idealist, Aristotle was a realist. Language was also important to Aristotle; to him, language was the tangible connection between a person's thoughts and the vocal expressions of truth. Everything that exists is made from the unity of form and matter. Only the supreme God represents pure form. Like Socrates and Plato, Aristotle believed that a reasoned, harmonious existence in moderation was the best way to live the good life.

The teachings of Aristotle encouraged two further schools of Greek philosophy, the Epicureans and the Stoics. The Epicureans believed the Gods paid little attention to humanity and that when we die, we simply cease to exist. There is no underworld or Hades. It is therefore best to live a life in moderation and in the pursuit of happiness. The Stoics believed that a spark of the divine lives in everything, an idea known as pantheism. Virtue and meaning were obtained by keeping true to your own inner resolve and purpose. The external world was volatile and cyclical; historical events constantly repeated themselves according to the movements of the heavens.

This golden age of Greek philosophy represented a new era in human consciousness. No longer was our existence dependent solely on the whims of the Gods or fate. For the first time in history, human reason became a focal point in determining how life should be lived and how we gain meaning from our material existence. Greek philosophy also gave birth to the possibility that human thought and reason could be linked with a kind of spiritual experience. Blind faith in Gods and tyrants was replaced by a new paradigm that still burns today: the idea that knowledge gained through an examination of empirical truth can enlighten each and all of us. As Christopher Hitchens famously put it, "philosophy begins where religion ends."[12] This period of time, paradoxically, also marked the first occasion that our instinct for spirituality could act as a tool for personal illumination about the material world without resorting to supernatural explanations. Sadly, this golden age was short lived.

Figure 16: Plato lecturing to a young Aristotle. This image is part of a larger painting by Raphael titled *The School of Athens,* which is on public display in the Vatican.

Religion by the Book

In ancient Rome, a different perspective and approach was taken with regard to managing individual spirituality and communal religious creations: the Romans put it in a box, separate and compartmentalised. In many ways, the practice of religion within both the Republic and the Empire represented a pagan paradise. There was a multitude of traditional Roman Gods. New Gods from all parts of the Empire could be incorporated into the Roman pantheon, as long as they did not interfere with traditional rituals or practices. All the Gods worshipped and venerated within the pantheon contained within them a divine spiritual essence called *numen*. This *numen* could be identified with special places, idols, and even humans themselves. For the Romans, their Gods were both intimate and specialised in form and function. There were household deities, family deities, and even Gods that ruled over physical items and structures. The Gods were worshipped communally by means of formal state-sanctioned temples and rituals. Priests were organised into colleges that performed ceremonial duties, divination, and sacred actions in a precise order of service. This emphasis on precision and accuracy became a distinctly Roman trait that not only applied to religious ritual and practice but also became embedded as a uniquely Roman way of getting things done.

Towards the end of the Republic, many of the elite and intellectual classes chose to identify with more abstract forms of religious worship that could be experienced through the mystery religions. These can be traced back to the philosophers of Classical Greece and even to the ancient mystery cults of Egypt. Greek philosophers such as Plato and Pythagoras were elevated as sons of Gods, as philosophy increasingly became intertwined with religious practice. In some Roman mystery cults, Plato was viewed as the son of Apollo, who, in turn, can be traced back to solar deities of ancient Egypt.

People's spiritual desires became focused on creating a sense of divinity that was increasingly human. In an attempt to stabilise the Empire, the first Roman emperor, Augustus (who ruled from 27 BCE to 14 CE), himself became the supreme priest of Rome, the Pontifex Maximus. This position, coupled with his role as emperor, gave Augustus a divine legitimacy; he was venerated as a kind of divine man and was worshipped for his genius. After his death, the Roman Senate formally confirmed his status as a God. The concept of a divine person who could influence and even save people originated with the teachings of the Classical Greek philosophers and evolved into a Roman search for divinely-inspired individuals who were chosen by the Gods as potential saviours of mankind.

Judaism and, later on, Christianity developed from a different perspective to that of the Romans, although both religious traditions were trapped within the confines of Roman rule. Instead, they combined the most sophisticated technology of their time – written language – and codified their religious beliefs. The God of the Hebrew Bible (the Old Testament) is without doubt a personal God, representing and even ruling the Jewish people. Monotheism, the belief in one supreme God, as expressed by the ancient Jewish people, represented a significant leap forward in how humanity envisaged the divine. This was a God that was in essence undefinable and beyond the comprehension of man. The ancient Jews did not have idols to worship; they would not even write or pronounce the name of their God.

Although this God was all-knowing and all-powerful, he was prepared to make a sacred covenant with his chosen people. In some ways, the God of the Hebrew Bible can be viewed as vengeful and violent. However, he does not demand that he be worshiped. The Jewish people choose him, and he chooses them – a covenant is made. The absence of any real indication of a significant afterlife in the earliest and most sacred book, the Torah (the first five books of the Old Testament), may indicate the dire need for the dispersed people of Israel to remain focused on the living covenant that they had with the Lord. Later Jewish literature contains inferences of an afterlife, either in heaven or in the new kingdom on Earth. The Hebrew Bible and the enduring faith of the Jewish people gave the Western religious tradition both a solid foundation and a cultural testament for living. A focus on written religious laws created a cultural meme that has endured for over two and a half thousand years. Judaism is a lived faith that has ensured tribal-based survival longer than any other Western religion. The price paid for this has been alienation, discrimination, and repeated genocide; but the sacred covenant remains intact.

It should be noted, however, that there was no one, uniform Jewish faith in the early years of the first millennium. There were two main Jewish sects that were popular to varying degrees at this time. First, the Sadducees were a priestly, aristocratic sect that emphasised adherence to Temple worship and sacrifice as per the strict written laws of the Torah. They did, however, embrace the concept of free will in human action under the domain of an all-knowing God. Second, the Pharisees were usually comprised of lesser Jewish nobles and priests. They believed in the premise of an eternal soul, spirits, angels, and the resurrection of the dead. The differences between these two sects are briefly mentioned in the New Testament. From Acts 23:8, "For the Sadducees claim that there is neither resurrection, nor angels nor spirits, while the Pharisees acknowledge all these things." Of the two sects, it was the Pharisees who had the most influence on what would become known as Rabbinic Judaism after the fall of the Second Temple. This is most probably because of their adherence to both oral and written law, which allowed some degree of doctrinal flexibility, as evolving oral law could reinterpret the meaning of written law. It is interesting to note that conceptual notions of free will and an immortal soul were (and are) contentious issues in Judaism.

Christianity was different. The new faith had two non-negotiable principles: it was exclusive – it did not tolerate other Gods or religious faiths; and it was evangelical – it wanted to expand its flock to all who would listen. The sect of early Christianity combined traditional Jewish teachings with the need to prepare for the coming Messiah through the individual display of love, compassion, and forgiveness. (In Judaism, the Messiah is an all-conquering warrior and liberator of the Jewish people). Jesus's main disciples, the apostles, began to spread his teachings throughout the land of Judea and also to the many Jews who were part of the diaspora across the Mediterranean. However, it was the writings and teachings of Saint Paul that gave a special light of faith for non-Jews. By the end of the first century, Christianity was a distinct and separate faith from traditional Judaism. The Christians were no longer a small sect but rather a rapidly growing religious faith.

Early Christianity, like Judaism, suffered under Roman rule. A convenient way to reduce civil unrest was to find someone to blame; this was often the Christians. They were an easy target. Not only were the Roman authorities suspicious of the secretive activities of Christian sects, but many of the Roman pagans were also wary of these strange people. Rumours about Christians engaging in cannibalism and incest abounded. From a Roman perspective, the Christians were seen as barbaric atheists because they refused to venerate the acknowledged Roman Gods.

Ironically, the more Christians were targeted for abuse, the more popular the Christian faith became in parts of the Empire. It was particularly popular with the slaves and the poor as it was essentially egalitarian and preached a gentle doctrine of love and forgiveness that could readily be accepted by those who had faced abuse and hardship. Jesus Christ himself had been tortured and put to death. The new faith also appealed to the intellectual classes as a monotheistic belief system that had credibility with its links to ancient Judaism. That said, both Judaism and Christianity were, and are, apocalyptic religions.

By the end of the fourth century, over half the people in the Roman Empire described themselves as Christian. Jews became increasingly marginalised, both in numbers and because they were perceived as the persecutors of Jesus. Christianity became the state-endorsed religion of Rome. Pagan sacrifice was banned and paganism itself declined in many areas under Roman rule. Jesus Christ became simultaneously fully human in his earthly existence and fully God in essence. The historical Jewish Jesus had been replaced by the redeeming Saviour of mankind.

Our genetically-inspired urge for spirituality was now under the watchful gaze and control of organised religion. Doctrine and dogma – culturally-created mechanisms – now held sway over any desire for individual spirituality. Pharaoh and emperor were replaced by an all-powerful Church ruled by a cabal of men. If our instinctual desire for spirituality is indeed an inheritable trait, most people throughout the history of Western civilisation have surrendered themselves to the totalitarian rule of self-proclaimed conduits of divine knowledge. Any 'spark of reason' would not be reignited until the revolution of the great Renaissance.

Figure 17: A fresco dating from the 16th century, depicting Roman Emperor Constantine and the delegation of Catholic bishops. Constantine was sympathetic to the Christian faith and even partially converted to Christianity after pleading to the Christian God for assistance in the major battle at Milvian Bridge in 312 CE. Subsequently, Christianity was made a legal faith in 313 CE. Constantine believed a unified Christian faith could help bring together the declining and fragmented Roman Empire.

Heresy

In the early centuries of the first millennium, a new spiritual paradigm emerged outside the confines of state-endorsed paganism or the dogmatic havens of Judaism and Christianity. The ancient Greeks had not only infused their world with the spark of reason, but they also inspired some individuals to later seek spiritual enlightenment without surrendering themselves to organised religion. For these people, their (genetically-sourced) desire to be spiritual beings was a personal journey.

Although still trapped within the confines of supernaturalism, this way of relating with a divine force was at least devoid of many of the control mechanisms enforced by organised institutional dogma and doctrine.

It is not surprising that these individuals were brutally persecuted. They believed that each person has within him or herself the promise of the divine. In other words, humanity has the potential to become like God. Redemption and salvation are journeys of self-discovery. The quest for illumination occurs outside the man-made structures of the Church, the state, and the monarchy. The critical element of this faith revolves around the goal of achieving personal freedom and gnosis (knowledge) about the true nature of the universe. Illumination is gained from secret gnosis. Followers of this path became known as Gnostics. They saw the Church as a disseminator of a hypocritical and false form of gnosis. The Church, on the other hand, believed the Gnostics represented the ultimate betrayal of Jesus Christ and his mission on Earth. In other words, the Gnostics were heretics!

Gnosticism was also sourced to earlier, secretive traditions where redemption, illumination, and immortality could be found through individual knowledge, which could only be accessed through the understanding of secret teachings. Therefore, redemption and immortality were not available to all people. Rather, they were restricted to those who had both the access to this secret knowledge and the ability to comprehend it. This form of secret teaching and revelation is known as esotericism and represents an antithesis to established Church doctrine.

The Catholic Church did everything possible to destroy the written works of the esoterics, and condemned them as heretics and enemies of the one true Catholic faith. The word 'esoteric' is derived from the ancient Greek word *esoteros,* meaning 'inner.' Western esotericism is a term that refers to a broad school of thought that, although it is historically linked to the Judeo-Christian tradition, stands apart with regard to its conceptualisation of God and the cosmos. There is no precise definition that can be applied to esotericism because the beliefs and practices of different schools of thought within the tradition vary greatly in both approach and perspective.

Some individuals also sought to reconcile philosophy, science, and religion; the most significant of these were the Neoplatonists. Neoplatonism evolved from the teachings of Plato and included works by various other Greek philosophers. Neoplatonists believed that although there was a supreme deity, this deity was indescribable, timeless, and immanent. For the Neoplatonists, gaining divine knowledge and grasping true reality were the ultimate purpose of human existence. Therefore, emphasis was placed not on physical being but on gaining knowledge that could transcend human analytical thinking. The physical or material world restricted humans to a lower way of thinking, one that obscured the absolute truths that could unlock the divine spark in all humans.

Neoplatonic ideas can also be traced back to Plato's allegory of the cave. In it, prisoners trapped in a cave can see only shadows as they appear on the cave wall. These shadows represent the only forms of being that can be seen by the prisoners. They cannot see the real forms that move behind the shadows and are further restricted from ever seeing the real world of light outside the cave. For the Neoplatonists, the material constraints of human reasoning are here illustrated by metaphor. In contrast, an enlightened person could communicate with the divine realm by gaining access to their own divine intellect or *nous*.

The main teacher of Neoplatonism was Plotinus (205–270 CE). Plotinus believed that most humans were at some level dissatisfied with their material existence and yearned for understanding and unity. In other words, Neoplatonists were instinctively searching for true meaning in life. They believed this could be achieved by a process of internal purification (catharsis) and contemplation in the deepest part of the mind. The ultimate aim of this process would be a unity of the cosmic trinity: the human soul, the mind, and the One. 'The One' constitutes the indefinable divine that both binds and unlocks all things. For Plotinus, the One was impersonal, gender-neutral, and oblivious to our human existence. This is unlike the God of the Gnostics and the Bible. It is by communing with the One that the soul and mind can transcend to the divine. "There we may see God and our self ... light itself, pure, buoyant, aerial, become – in truth, being a – God."[13]

The fall of Rome in 476 CE meant more than just the collapse of a military empire; it was also the catalyst for the decimation of a uniquely Western way of viewing the world and the loss of a thousand years of scientific and cultural achievements. Ways of seeing the world through the lens of individually-sourced spirituality also diminished. People now either surrendered themselves to mother Church, or they embraced other forms of magical supernaturalism. What had commenced over a thousand years earlier as a quest for understanding ourselves and the natural world, was again a tyrannical control mechanism that ensured the survival of a ruling elite. Our personal yearning for spirituality had culturally-evolved to serve organised religion as slaves to an all-knowing Catholic Church and her monarchical representatives. Any hope of spiritual rebellion inspired by the Gnostics, Neoplatonists, or other seekers of personal enlightenment were brutally crushed. Yet another uncomfortable truth emerged; a dark age descended on humanity, and it would last a thousand years.

Figure 18: Sophia (Greek for 'wisdom'). Both the Gnostic and Christian traditions venerated Sophia as the divine personification of wisdom. This link back to ancient Greek philosophy – the love of wisdom – shows that some early believers may have sought rational truths about the world around them, while still subservient to supernaturalism.

Renaissance and Enlightenment

The final gasps of what remained of any semblance of Roman rule came to an end with the fall of Constantinople in 1453 to the Ottoman Turks. Many Greek-speaking intellectuals who had resided in the city were dispersed to safer parts of Europe and north Africa. They took with them both their Greek heritage and what was left of ancient Greco-Roman literature. The translation of these texts gave a direct pathway to the Classical past that was immediately embraced by the leaders of the Italian Renaissance. Intellectuals who were interested in learning from these past great civilisations could now gain access to knowledge from a golden age that emphasised a rational, humanist approach to viewing the world around them. As a result, new cultural interpretations were reflected in the writing, art, sculpture, science, music, religion, and philosophy of the Renaissance.

In creating these new visions and interpretations, people looked to the ancient past as the basis for truth and beauty. The Renaissance recaptured the spirit of those times, its celebration of the intellect and belief in the innate ability of people to influence and interpret the world around them. Interestingly, the Renaissance would produce two contrasting forces that provided conflicting conclusions on the question of how things came to be as they are. First, there was a return to a celebration of the role and place of the individual as a unique and powerful being. This freedom of thought also inspired later Renaissance thinkers to create a scientific revolution, where the search for truth through mathematics, science, and empirical enquiry would eventually lead to the rejection of many Classical concepts. This was especially so with regards to the long-held view of the earth and humanity as being at the centre of God's universe.

In a Classical approach to enquiry from the Greek tradition, emphasis was placed on understanding absolute truths about the universe and the idea that these truths could be deduced by certain enlightened individuals. The new scientific revolution, however, placed emphasis on not only discovering but also controlling the natural world. Knowledge was gained by understanding as well as through continued observations of how natural things worked and how they could be used. But the acceptance of scientific and rational ways observing the world came at a cost. For example, physician Michael Servetus (1510–1553) – the first person to observe that blood flows from the heart to the lungs and back – was burned for doubting the divine Trinity.

In 1543, Nicolas Copernicus, a Polish scientist and philosopher, published *On the Revolution of the Celestial Orbs*. Copernicus argued that the sun, not the earth, was the centre of the universe – or, more precisely, of the solar system. Ironically, the ancient Greek philosopher Archimedes had expressed the same opinion two thousand years before. In this heliocentric view of the universe, the planets made fixed, uniform, circular rotations around the sun. This revolutionary work by Copernicus was not deemed heretical, even though it displaced the earth as the centre of all things, and was even recommended for wider distribution by the Pope.

This new theory of cosmology was later refined by Johannes Kepler, who successfully argued in 1609 that the orbits of the planets in our solar system were not circular but rather moved around the sun in the form of an ellipse. Even the moon's orbit around the earth was that of an ellipse. The premise that the planets revolved around the sun elliptically meant that some other force was impacting on these movements. This idea would inspire scientists such as Isaac Newton to discover the universal laws of gravity and movement. Galileo gave visual verification to the theories of Copernicus and Kepler and gained the full attention of the Catholic Church, which immediately declared that a heliocentric view of the universe was now a heresy. Galileo not only published these truths in the *Dialogue Concerning the Two Chief Systems* (1632) and *Discourses and Mathematical Demonstrations Relating to Two New Sciences* (1638), he also satirised the Pope.

Given that the heliocentric view of the universe was now deemed heretical, Galileo was engaging in very risky rhetoric. He went further by denouncing some of the writings of the Old Testament as "written by the ignorant for the ignorant." He spent the final eight years of his life imprisoned in his home on the direct authority of Rome. For the first time, empirical, progressive science had more credibility than religious constructions regarding cosmology. People could look up at the stars and see the wonders of science, not just the glory of God. The new science also looked to the future, not back to a glorious Classical past, and was finding new ways of reasoning through a progressive spiral of accumulating knowledge.

Again, however, some enlightened individuals paid the supreme price for expressing reasoned truth. In Italy, renowned scientist Bruno Giordano (1548–1600) was burned at the stake by the Holy Inquisition for teaching that the earth circles the sun. Even given the precariousness of the times, people like

Rene Descartes and Francis Bacon sought to form a rigorous methodology of scientific investigation. Descartes, in his most famous work, *Discourse on Method,* "sought an axiomatic method of proof capable of both establishing incontrovertible propositions and defeating radical skepticism."[14] In other words, he argued that scientific research should be objective and considered consistent if it was devoid of contradiction. Descartes even ventured into the sacred realm of what would later be termed the philosophy of the mind. Famously known for the statement "I think, therefore I am," he reasoned that personal reality exists separate to materiality. Although both Rene Descartes and Francis Bacon were shining lights of a new enlightenment, they appeared to hold firm to their religious convictions, albeit from a very liberal perspective.

Opposition to the doctrinal rule of the Catholic Church was not limited to arguments over science and religion; a civil war now exploded from within: the Reformation. Begun by Martin Luther, new Protestant religions soon expanded across Europe. This revolution in belief would inspire diverse branches of Protestantism, including the Methodists, Presbyterians, Baptists, Anabaptists, Puritans, Quakers, and other alternative forms of Christianity. The Reformation came at great cost in terms of human suffering, through both persecution and wars between the faiths. Germany lost nearly 40 percent of its population during the Thirty Years' War (1618–1648). This war between Protestants and Catholics effectively reconstructed the map of Europe.

The period spanning the Renaissance and the Reformation was both a time of the resurgence of older spiritual traditions as well as new interpretations of Christian faith. The

Roman Catholic Church lost its monopoly on Christian doctrine in Western Europe and also lost much of its political influence. The teachings of John Calvin inspired the Puritan revolution in England. Puritan Protestants were among the first white settlers in the New World of North America. The Calvinist ethic of acting as though they were chosen by God in some ways influenced the American idea of being a "chosen nation whose flag and ideals have a semi divine purpose."[15] The Catholic Church was now opposed from a variety of fronts. Protestant reformers wanted a faith that was grounded in a belief structure that used the Bible as its only true scripture. This ideal would lead to internal conflict among different Protestant faiths with respect to how to interpret the words of the Bible. Those people, such as the esoterics, who wanted a more personal relationship with the divine also abandoned the old Church and searched for meaning in other, more secretive ways.

For some people, the gradual awakening of empirical scientific practice and research were seen as complementary to esoteric beliefs and ideas. Science showed how things worked in the material world, and esoteric teachings provided a tangible connection between the material and the unknown divine. Even the greatest scientific mind of post-Reformation Renaissance society, Sir Isaac Newton, was a practicing alchemist. He became known as the last great magician who sought to "Discern different intelligences, perhaps even distinct principles with distinct centres of consciousness behind the material surface of things."[16]

If one cultural meme endured from the revolutionary turmoil of the Renaissance, Reformation, and subsequent Enlightenment, it was the message that organised religion – in particular, the monopoly of power over what could be defined as truth disseminated by the Roman Catholic Church – no longer answered all things for all people. Our genetically-derived urge to be spiritual beings now expressed itself in a variety of ways; some people stayed in the arms of the Catholic Church, some 'protested' by creating new faiths or engaging in esoteric practises, and some tried to reconcile their desire to be spiritual with the reality of scientific truth. If, as evolutionary psychologists argue, each of us is born with a genetically-programmed module which is disposed to seeking spiritual fulfilment, then a dilemma now existed as to how this desire could be satisfied if structured religion was disregarded. This remains one of the true existential crises faced by many of us today.

Figure 19: An artistic representation of the heliocentric nature of the solar system. Note the elliptical path of the planets around the sun. The discoveries revealed by science increasingly proved that our world, and even the so-called heavens, were ruled by natural law, not divine decree.

Key Points

- Our innate and anciently-sourced desire to be spiritual encouraged some individuals – usually a tribal priest/shaman – to observe the workings of the world around them. But it was the ancient Greeks who were the first civilisation to look inwards at themselves to find meaning and understanding.

- Humanity, through the use of reasoned thought and action, could find a pathway to comprehending the material world.

- Greek philosophy also gave birth to the possibility that human thought and reason could be linked with a kind of spiritual experience.

- The ancient Greeks were also the first Western civilisation to recognise that the culture created to nurture communal living was the only thing that restrained the chaotic nature of individual citizens.

- It also became clear that when structured religion is sidelined, civilisation flourishes.

- In ancient Rome, the workings of the state were, in the most part, separate from everyday religious practise. However, towards the end of Roman rule, the exclusive and evangelical rule of written dogma and doctrine that has always typified Christianity would hold sway on all aspects of Western life, and would continue to for a thousand years.

- The Renaissance and the Reformation allowed the resurgence of older spiritual traditions as well as new interpretations of Christian faith.

- A new scientific revolution emerged which placed emphasis on not only discovering but also controlling the natural world. Knowledge was gained by understanding as well as through continued observations of how natural things worked and how they could be used.

- Although many of the pioneers of scientific rationalism still held spiritual beliefs, there was no reconciliation between science and structured religion – they were diametrically opposed. People now had a clear choice; either surrender themselves to the supernatural will of God as dictated by the words of their chosen religion, or scientifically search for rational truths about the world in which they lived.

SEVEN

A Reckoning

Atheistic existentialism, of which I am a representative, declares with greater consistency that if God does not exist there is at least one being whose existence comes before its essence, a being which exists before it can be defined by any conception of it. That being is man.

John Paul Sartre, *Existentialism is a Humanism*

Introduction

By the time of Charles Darwin, Western religion had lost the power to kill heretics and other nonconformists. New scientific discoveries opened up an emerging reality that chipped away the supernatural explanations of religiously-derived truths. For many people, religion and science now became open adversaries in a battle that still rages today. A notorious example of this was the 'Scopes Monkey Trial' in Tennessee in 1925, where Christian fundamentalists attempted to ban evolution from public school science courses. The fundamentalists contended that a supernatural God made all species in modern form about 6,000 years ago, while science backed the premise of evolution by natural section over very long periods of time. The defendant, high school science teacher John Scopes, was found guilty and fined $100. Both sides claimed victory.

The emergence of democratic capitalism after World War II as the dominant form of Western economic structure, was typified by a society that made economic growth the prime indicator of a life well lived. The second half of the 20th century also became known as the century of the self. Paradoxically, many new personal desires were themselves created through means such as advertising, media manipulation, public policy, and reformed religious doctrine. The ruling elites of organised religion, business, and government had simply adapted their manipulation of the masses by indirect means. The main premise behind democratic capitalism as practised in the United States and other Western countries was that the individual consumer was the cornerstone of society. This was known in economic terms as consumer sovereignty. However, the power of individual choice was either an illusion or stage-managed by outside interests that manipulated the individuality and behaviour of citizens.

Advertising created a veneer of personal choice and individuality, but consumer sovereignty was subordinated to producer sovereignty, patriotic nationalism, and organised religion. It could be argued that, given the violent and chaotic history of humanity, control over the individual and his or her beliefs was necessary in any society that wanted to maintain peace and order. The price paid for this control, however, is the handing over of power to other individuals within the controlling structures who have the same inherently selfish genes.

If our cultural constructions are to play an important role in the formation of a new spiritual and moral paradigm, then we need to stand openly and honestly at the world we have created. The 20th century was the most violent and turbulent period of time in the history of Western civilisation. The earth and its peoples were scarred by two world wars, a deep economic depression, the decimation of the natural environment, genocide, and the invention of weapons that could destroy the planet. It became clear that advanced science could also be used to dominate and destroy much in the same way that dogma and doctrine served the desires of totalitarian elites of the past. Some people even asked themselves whether the structures that held civilisation together were themselves absurd or even evil creations. Achievements in art, literature, and culture were outweighed by the absolute and deliberate destruction of many aspects of established life. If our cultural institutions are the primary hope in civilising our selfish genetic nature, then how we govern ourselves and how we create positive rules of moral behaviour should also be closely examined.

The 20th century also represented a turning point in Western approaches to the question of how to live a good life. Capitalist materialism dictated loyalty to the traditional structures of the Church and state as the defenders of personal faith and freedom. An individual could live out his or her life as a unique and free entity as long as they acknowledged the higher communal roles of government and Church. But towards the end of the century, there appeared to be a perfect storm of discontent forming in the hearts and minds of many people who felt alienated from the industrial world that had been constructed around them. They no longer trusted their leaders and the doctrines and practices of the major religions in the search for a meaningful life. Alternative paths, including the search for meaning beyond the material rewards of capitalism and consumerism, began to appeal to people who were burdened with a crippling sense of disillusionment. This sense of personal alienation, coupled with environmental and cultural disintegration, is the reality that now presents itself in the 21st century.

An Existential Crisis

The 20th century was also a time of introspection about what it meant to be human and how a person could experience a life well lived within the confines of their own personal individuality. Alternative ways of knowing and experiencing a meaningful life gained traction in some parts of Western society. Probably the most significant of these alternative paths was that of the existentialists. The existentialists believed that the construction of a meaningful life is the personal responsibility of each individual.

For the existentialists, a person could now create their own sense of being and meaning outside the confines of institutional structures such as the Church and the state. Life was a personal journey bounded by the unique individuality of each person. The creations of the state and organised religion were essentially absurd. The main exponent of existentialism, John Paul Sartre, wrote that "Man is nothing else but what he makes of himself."[1] This one sentence sums up the rejection and frustration of a generation that had suffered under the corrupt structures of a government and Church that had decimated the world around them.

Existentialism was itself influenced by the earlier philosophical school of nihilism. Nihilism contends that human existence is an insignificant component in an indifferent and chaotic universe. Institutions such as the Church and the state are useless structures that are incapable of providing meaning. As such, each person is absolutely alone with his or her terrifying consciousness of both mortality and insignificance. As renowned German philosopher Martin Heidegger says, "Estranged from the community of being in one whole, his consciousness only makes him a foreigner in the world, and in every act of true reflection tells of this stark foreignness."[2] This miserable and bleak outlook on the ontology of human existence led to the emergence of existentialism to counterbalance the dismal stance of nihilism.

One of the great paradoxes concerning human existence is the reality that while our social and cultural constructions have evolved to dampen some of the destructive impulses individuals are genetically bequeathed, these same institutions appear to also hinder each person's personal quest for inner enlightenment. If we also take away the blindfold of religious faith and the cane of religious doctrine, we stand alone – either empowered to create our own meaning, or stranded in a meaningless void.

Albert Camus expressed the innate discontent that existed in the hearts of people who believed the world around them was an absurd prison. In his essay, "The Myth of Sisyphus," he describes the ancient Greek story of Sisyphus, doomed to live out his life pushing a boulder up a mountain only to see it fall back down. Camus reconceptualises this miserable existence using the idea that Sisyphus actually gains a sense of personal meaning from merely doing the task and participating in the human struggle for existence. In other words, Sisyphus creates his own sense of worth and purpose.

Camus himself never claimed to belong to the existentialist school of philosophical thought. In later works such as *The Outsider*, he appears to confirm the dominance of inherent absurdity and nihilism over that of existentialism. However, Camus directly addressed the meaningless void of absurdity in a way that allowed later existentialists to create a paradigm that gave hope in the face of nothingness.

The emergence of existentialism coincided with rapid generational changes in Western society. The 1960s, in particular, were a period of rebellion by young people and involved a search for new ways of viewing the world, either through alternative social structures or by escaping into chemically-induced altered states of being. For many people in the later part of the 20th century, the pursuit of individual happiness and fulfilment were outside the boundaries of structured civilised society – eventually, each would oppose the other.

The counterculture of the 1960s was primarily a reaction to the controlled capitalism of post-war societies that valued conformity and the structured values of Church and state. Young people in the United States and Australia had further impetus to rebel against authority. The draft imposed by their governments put a whole generation of youth in the firing line that was the Vietnam War. The subsequent growth of political and social subcultures, such as the anti-war movement, not only influenced changes in government policy but also created an environment in which the individual became the focal point of living a meaningful life.

Institutions such as Christian churches (in all denominations) and governments (in all their manifestations) were no longer trusted to either provide honest guidance to citizens or maintain a peaceful environment that nurtured the spiritual needs of their people. An increasing number of individuals sought alternative ways in which to transcend the misery and boredom of everyday existence. This tumultuous time may have had its genesis in the teachings of the existentialist philosophers who believed that forging a purposeful life was the individual responsibility of each person. The importance of personal inner journeys for enlightenment overshadowed any sense of communal responsibility. It is not surprising that the new drug of choice for this generation was lysergic acid diethylamide (LSD).

LSD was a mind-expanding and mind-altering drug that promised to not only open new doors in terms of personal reality, but also to answer deeper questions about how a person experienced the intangible link between the spiritual world and the material domain. Author and philosopher Aldous Huxley pioneered the exploration of altered states of mind as paths to a higher form of existence. In his book *The Doors of Perception*, Huxley advocates the use of the drug mescalin as a means of transcending reality. He provides the reader with vivid descriptions of his own 'trip' to another world. Mescalin is a naturally-occurring drug that can be found in the peyote cactus. It has been used for thousands of years by native South Americans to enter altered states of mind similar to those produced by the ingestion of LSD.

Huxley described in detail the historical desire of humanity to escape the confines of everyday life. For Huxley, humanity has an innate desire for something more, that is, for an understanding of its role and purpose in the universe. Huxley says:

But the man who comes back through the Door in the Wall will never be quite the same as the man who went out. He will be wiser but less cocksure, happier but less self-satisfied, humbler in acknowledging his ignorance yet better equipped to understand the relationship of words to things of systematic reasoning to the unfathomable Mystery which it tries, forever vainly, to comprehend.[3]

Renowned psychiatrist Carl Jung contended that similar ritualistic practices across time and culture could reflect universal archetypal symbols that are usually hidden from conscious experience. For example, Jung contended that the ancient mother Goddess is an ongoing archetypal image that is deeply embedded in the collective unconscious of humanity. This archetype lends us protection, fertility, and a tangible link to the natural world. These symbolic representations create a sense of immortality of spirit through a shared memory of experience. They are not supernatural beings, but rather the legacy of our genetic past and our communal yearning for spirituality. Therefore, any journey into altered states will result in the emergence of archetypal universal symbols that exist in all of us.

During the final years of his life, Jung admitted the superiority of the inner person to that of external reality.[4] Jung was a spiritual existentialist who rejected the confines of organised religion and belief in a personal God. In contrast he placed his faith in the individual (and communal) ability of humanity to spiritually transcend the confines of materiality without bending our knees. As Jung put it, "There are, and always have been, those who cannot help but see that the world and its experiences are in the nature of a symbol, and that it really reflects something that lies hidden in the subject himself, in his own transubjective reality."[5] That said, it should also be recognised that some of Jung's more abstract observations – particularly those concerned with the idea of synchronicity – border on occult ways of knowing. Although thinkers such as Jung and Huxley sought mystical (even supernatural) solutions to the existential dilemmas facing us, they at least acknowledged the need for personal empowerment and responsibility in living a meaningful life outside the prison of any structured religion.

By the end of the 20th century, there appeared to be yet another perfect storm of discontent forming in the hearts and minds of many people who felt alienated from the industrial world that had been constructed around them. They also no longer trusted their leaders and the doctrines and practices of the major religions in the search for a meaningful life. Alternative paths, including the search for spirituality beyond the material rewards of capitalism and consumerism, began to appeal to people who were burdened with a crippling sense of disillusionment. Retrospectively, it appears that existential philosophers sensed that something inside us instinctively controls how we see the world around us and how we live our lives. Although they did not emphasize (or even recognise) the role of evolutionary biology or psychology, they at least recognised that there was something inherently missing in ourselves and in our institutional creations – that the way 'we are' is not the way we 'should be.'

Figure 20: *Sisyphus* by Titian (1549). Many workers in highly industrialised societies would identify with the curse imposed on Sisyphus. Without the comforting blindfold of religious faith, we are left alone to search for spiritual meaning.

A New Search For Meaning

Humanity is the only life form that is completely aware of its own mortality. The events of the 20th century, and the accelerated chaos that is 21st century life, have complicated and deepened the dilemma of our own personal demise. We now, as a civilisation, have the capacity to destroy every living creature by means of a man-made nuclear Armageddon. All life on this planet could be destroyed forever. Paradoxically, recent decades have also witnessed exponential advances in technology and medicine; even conquering the diseases of old age is no longer seen as impossible or implausible. Another challenge, however, remains: how do we see ourselves, others, and the world around us? It is in this period of time – our time – that a battle of beliefs becomes once more the focal point in the spiritual discourse of the Western world.

The 21st century has already revealed that it will be a period in history where the advancements of science and technology will impact directly on the evolutionary advancement of humanity. How religious institutions relate with new visions of the human form will also be an interesting challenge. Christianity, the main religion in most Western countries, dictates that man was created in the image of his God, not the machine. Life itself was something to be endured. The faithful will be rewarded when they die, not when they merge with the machine. That said, the possible abandonment of religious supernaturalism does not mean the abandonment of spirituality. Our imagination, sense of truth and beauty, compassion, and our innate desire to seek understanding of the world around us, may play an even more important role in creating a meaningful life – there may indeed be 'more than this.'

It must, however, be conceded that many people choose to reject the sensual experiences of material existence. They now prefer to engage with others in a virtual world of constructed reality. Social media and other cultural expressions are increasingly mediated by a technocracy made up of a few organisations that dominate the communication of information. This has become the dominant means of searching for meaning at the beginning of the third millennium. There has also been a significant shift not only in the accessibility and variety of information available to people, but in how people send and receive pieces of information. It has only been in the last century that mass media, beginning with radio and television, gave people instant access to culture, news, and events. These communication media required the receiver to be a passive participant in the process. The advent of dynamic meta-information in the form of interactive online media has made each person a player in the field of communal information. Individuals now have an instant power of reply. Traditional forms of media have diminished in influence to the extent that the newspaper, radio, and network television are almost redundant as communication devices.

However, while the power of opinion has grown, and while almost anyone who wishes to voice anything can now do so and be heard all over the globe, that opinion does mostly originate from the privacy of the home. An individual participates globally while within the intimate confines of their bedroom or lounge room and perhaps while in their pyjamas. There has been a diminishing of face-to-face modes of communication. When in public, people often prefer to talk to their mobile device rather than engage in conversation. From a religious perspective, this new preference for communication that is mediated by technology diminishes the importance of communal religiosity, that is, the desire to religiously engage with a community of like-minded people in the real world. The celebration of the Christian Mass in a congregation of believers is foreign to most people, who might prefer to interact with their peers in the global village. This poses an enormous challenge to all forms of structured religious faiths, which have always posited a community of believers held within the arms of a physical and tangible place of worship. Less than ten percent of people in Western countries attend traditional church services, even though over 70 percent of them still believe in God.

That said, our personal efforts so far to either escape or transcend the material world have, at best, yielded mixed results, but it does appear that some of us are now prepared to stand at the abyss without a supernatural safety net and see ourselves for who we really are – or who we really want to be. It does not matter if this reality is augmented and/or supported by technology. Most of us can now access information and communicate outside the control of institutional dogma and doctrine. We can now independently choose how we exercise our individual and communal desire to reach for something more than this – and for those brave enough, it will not be the hand of God. I believe this will be the enduring legacy bequeathed to us by the four horsemen: Dawkins, Dennett, Hitchens, and Harris.

Sadly, it still must be conceded, however, that many otherwise intelligent, rational people continue to openly embrace religious belief in a personal God. They and their families go to church, pray and hope to be saved so that they can enjoy the eternal bliss of heaven. If one cultural meme has endured for thousands of years, it is the hope for a heavenly (or at least other-worldly) afterlife. Our heritable yearning to be spiritual beings has been a perfect nesting place for religious dogma and doctrine. Indoctrinated since childhood, the faithful can live a fulfilled life within the arms of mother Church under the love and protection of a heavenly father. However, this comfortable delusion falls apart when the explanations of science break down the walls of supernatural belief, where the ornaments and preaching of religious leaders are exposed as human creations.

If one thing emerges clearly from the fog of history, it is the fact that organised religion struggles when individuals search for meaning either from within themselves or from reasoned analysis of the world around them. This struggle denigrates into aggressive conflict when the truths of science are introduced to destroy the validity of religious doctrine and dogma. This occurred in ancient Greece, in parts of the Roman Empire, and during the great Renaissance. Taken one step further, it is also apparent that the real conflict is one of either submission or illumination. We can choose to cling onto man-made constructions of communal faith in a supernatural being that has the power to selectively save or punish his followers, or instead choose to seek an internal and personal search for illumination. The reality of our genetic tendency to be spiritual beings makes this choice compulsory. Either join the congregation and put your faith in God and his Church; or take a personal journey inward which is mediated by each person through his or her own thoughts and actions.

Existentially, all we really have is ourselves and our intrinsic bond to humanity. Any quest of illumination is personal, and therefore, it is ultimately one that is taken alone. This premise is obviously frightening to people who believe in a personal God who acts as a protector and guide in life, as well as a pathway to paradise in death. I will contend in the following chapter that there are other forces at play today which are attempting to create a revolution in the way we view ourselves and our purpose in the scheme of all things—human and spiritual.

Figure 21: In 1927, German film director Fritz Lang released the science fiction film *Metropolis*. In it, Lang gives us his dystopic vision of the world in 2027. Humanity is divided by class, and technology is used by an elite cabal of men to enslave others. The film's heroine, Maria, is kidnapped, and her likeness is transformed into a robot who unleashes chaos over the city. Lang bravely posed the questions: what do we want to become, and what type of world do we wish to exist in?

Key Points

- For all the benefits bequeathed to us by the emerging dominance of science and our societal creations, they cannot create personal inner meaning. Our cultural institutions may even hinder any quest for inner enlightenment.

- If we also disregard structured religion, we stand alone – either empowered to create our own meaning, or stranded in a meaningless void. Existential philosophers sensed that something inside us instinctively controls how we see the world around us and how we live our lives.

- They also recognised that there was something inherently missing in ourselves and in our institutional creations – the way 'we are' is not the way we 'should be.'

- If our cultural and spiritual constructions are to play an important role in the formation of a new spiritual and moral paradigm, then we need to look openly and honestly at not only the world we have created, but also at how we have created our own inner life journey.

- Organised religion dictates how a person should live their life and therefore becomes irrelevant when individuals search for meaning either from within themselves, or from reasoned analysis of the world around them.

- The 21st century has already revealed that it will be a period in history where the advancements of science and technology will impact directly on the evolutionary and social advancement of humanity at the expense of structured religious belief. But the abandonment of religious supernaturalism does not mean the abandonment of spirituality.

- Our imagination, sense of truth and beauty, compassion, and innate desire to seek understanding of the world around us may play an even more important role in creating a meaningful life.

- New realities are emerging that will slingshot humanity into ways of being that will be augmented and/or supported by technology. For the first time, we may now be able to stand at the abyss without a supernatural safety net and see ourselves for who we really are – or who we really want to be.

EIGHT

An Enduring Spiritual Meme

After Darwin, God's role changes from being the designer of all creatures great and small to being the designer of the laws of nature, from which natural selection can unfold, to being perhaps just the chooser of the laws. By the time God's role has been so diminished, he becomes a bit like a constitutional monarch, presiding ceremonially but not having any more work to do.

Daniel Dennett

Introduction

There have been critical turning points in history when religious doctrines and practices did not adapt to a dynamic and changing real world. As previously discussed, an example of this would be the clash of beliefs that occurred during the Renaissance and the Enlightenment. The revelations of scientific thought and the idea that the truth could be deduced by reason meant that many people no longer accepted their chosen religion as the sole arbiter of how to live a just and good life. it is increasingly apparent that structured religion stands diametrically opposed to the truths given to us by science. The religious surrender themselves to faith, thus negating any need to justify their actions rationally. Reasoned argument between the two cannot exist: religion is dependent upon supernaturalism, while science rests completely on the laws of nature. Most Western religious traditions have also been (and are) based on the idea that the world is dominated by opposing supernatural forces of good and evil, a divinely-designed conflict culminating in Armageddon. Dualism, as it exists from a scientific perspective, represents opposing positions that can be identified and confirmed (or falsified) through objective observation and/or through reasoned logic.

The concept of dualism has also been expressed in the division between body and soul, mind and soul, and materiality as opposed to spirituality. Sigmund Freud was only too aware of the duality which exists within a person's mind. His concept of the id, our instinctual drive for gratification, and the subsequent development of ego and super ego as mechanisms to suppress the id, originated before the significant discoveries of genetic biology and evolutionary psychology. As previously noted, science now concedes that human behaviour, as well as our physical traits, are influenced by our genes. Any notions of so-called evil are therefore embedded in genes which have survived hundreds of thousands of years of relentless natural selection.

Richard Dawkins argues further that it is impossible to obtain any moral values from natural selection, as "the process is vicious, brutal and short-sighted."[1] Therefore, any hope for the communal 'good' must originate in those groupings of individuals who can envision a future outside of selfish genetic determinism. As evolutionary biologist T.H. Huxley concludes, "If you must use Darwinism as a morality play, it is an awful warning. Nature really is red in tooth and claw. The weakest really do go to the wall, and natural selection really does favour selfish genes."[2]

Conversely, I also concur with the idea that any ethical or personal spiritual framework that best suits the challenging world we live in today, must disregard any notions that religion creates and defends our communal sense of morality. To reference the moral laws given to us in archaic religious texts is not a solution; it is a disturbing distraction and a selective delusion. At best these laws can give us a social reference point for re-evaluating our collective religiosity over time. Therefore, it must be acknowledged that we are now entrapped in an ethical vacuum, where we must not only recognise our selfish and unstable nature, but can also no longer afford the comfortable delusion that religion is the source of our moral code. This premise is not new; anyone who honestly observes the reality of our recent history stares into an abyss of mass delusion and chaos. Is it possible to reconcile our yearning to be spiritual beings with a moral framework that aspires to be universal while empowering individual agency for the good of all?

This rebellion against our primitive and inherently selfish nature may give us a form of altruism which could be embraced communally. If traditional religious institutions gave up on any notions of exclusivity, supernaturalism, and divinely-inspired law, then their more rational and compassionate beliefs could be a conduit for 'super niceness' within and between cultures. Richard Holloway, the former bishop of Edinburgh (who now calls himself a recovering Christian), believes modern humanity represents a "singularity in evolution." We are the first known living species to transcend the chains of our nature as bequeathed to us by our genes – we now have the ability to predict and to some extent foresee our future. Ironically, our desire to be spiritual – the trait used so many times for mass manipulation – may be a unifying force to reach for something 'more than this.' In other words, it may be time for scientifically-inspired spirituality.

Finding a New Way

The last words spoken by Jesus are found in Revelation 22:20: "I am coming soon." These words would have undoubtedly comforted many early Christians who were poor and abused by the world around them. However, the end of days of Paul and Peter have not eventuated. There has been no divine Armageddon or judgment of the risen dead. Instead, humanity has continued its erratic and violent journey into an unknown future. The challenge for any faith tradition in the 21st century is to claim the best of the communal aspects of our spiritual journey, and to infuse into this the personal quest for individual enlightenment. The idea of a flesh-and-blood resurrection of all humanity 2,000 years after the death of Jesus Christ places Christian faith outside the realm of even mythical truth. Myth has been transformed into fiction because it no longer fits the hyper-real paradigm of today's society, which is suspicious of institutional creations, including our religions, and even, to some extent, the historical legacy of Western civilisation.

That said, it must be conceded that many Western societies have unfortunately experienced a dramatic increase in the number of people embracing a 'fundamental' stance on their chosen religious faith. Fundamentalists take their scripture at face value and reject any other interpretations of holy dogma, scripture, or rational truth that does not fit with their beliefs. Like the early Christians, fundamentalists are also evangelical and zealous in spreading their faith to all who will listen. For these people, scientific truth is at best supplementary to scripture, and at worst the work of the Devil, designed to corrupt humanity. But it could also be argued that all forms of structured religion represent a 'fundamentalist' perspective on the world. As mentioned previously, any belief in a supernatural and personal God who favours only his faithful is outside the realm of rational truth and morality. Therefore, attempts by so-called moderate believers to de-radicalise their fundamentalist brothers and sisters is akin to a drug addict substituting heroin for morphine — both parties remain addicted to their faith as the exclusive path to salvation and paradise in heaven.

Structured religions and beliefs in a personal God may have played a bonding and control mechanism to ignite, suppress, or redirect our violent nature, but they are no longer relevant or meaningful in a new world where the truths of empirical science and reasoned thought pay no heed to archaic books of blind faith. Daniel Dennett believes that one of the main reasons religions are failing is due to the democratisation of knowledge. He extends this premise to recommend "non-ideological education for boys and girls in every community on the globe."[3] We can now independently find truths about ourselves and the world around us.

However, the democratisation of knowledge is a two-edged sword. Atheism, if it is to be taken in its literal form, insists that humanity is just another organism that lives and dies according to natural law. For atheists, there is no predetermined plan that sees the universe as an intelligent creation. If we are indeed born with a yearning to be spiritual beings, then a belief in atheism may not only be frightening for many people, but it may also be against our inherited nature. A deeper question now arises: why is this trait so hard-wired within us, and how can we culturally adapt it to work as a force for the 'rational good'? Personally, I would like to think that there is some greater narrative to the evolution and history of our species. I understand that it may sound grandiose to assume that humanity has been (for want of a better word) chosen to transcend the confines of the material world and conquer the natural laws that govern us, but this is precisely what has transpired.

Our ability to do this is not some gift from God, but rather the result our ability to culturally and, more recently, scientifically intervene in the workings of natural selection. We are, by nature, intelligent and spiritual – this salient fact must be recognised as an intrinsic part of who we are. Only intelligent and personally empowered spiritual beings will be capable of creating a new world which embraces our technological achievements without discarding the more noble aspects of our species or the chaotic beauty of the planet we live in. I believe that if we cannot rise above the chains of fundamentalism and the innate misery of nihilistic atheism, then we are doomed to subservience.

Figure 22: Leonardo da Vinci's *Vitruvian Man* as depicted in a notebook around 1490. Da Vinci gave humanist proof to the perfection of humanity in the outstretched image of a man inside a perfect circle, bounded by a perfect square. Note that the centre of the circle is the navel, and the centre of the square is focused on the genital region. This gives a perfect pictorial representation of the interdependence of the spiritual and the material within the human form.

Spirituality Without God

More people are now grasping for a form of spirituality that can provide personal hope and meaning without compromising the hard-earned truths science has bequeathed to us. They yearn to be spiritual without supernaturalism. Although some individuals have dared to look into this cold reality, many of us still reach blindly for the love of God. Today, it is estimated that humanity has produced over 100,000 religions, and at least that many Gods. (A brief history of deity incarnation from a uniquely Western perspective can be found in Appendix 1.) A critical question a person could first ask themselves is this: What type of God do they believe in? An atheist would reply that there is no God. A follower of any of the Western monotheistic religions could reply: Our God created humanity and everything in the universe. He is caring, loving, and listens to our prayers and appreciates our loyalty and fidelity to him, and is prepared to supernaturally intervene on our behalf.

The person wishing with all their heart that God will help the roulette ball land on lucky seven has the same faith perspective as a person diligently praying for a cure to a disease inflicted on a loved one. They are both pleading for a suspension of the laws of nature. They hope God will favour them and grant their wishes and prayers. They are not alone. Their God listens to them, protects them, and guides them through the struggles of life. A simpler but broader definition of this God could be: a supreme, universal, and divine force that selectively and miraculously rules over all things. God is perfect. Conversely, a deist would answer: Yes, I do believe that God created everything that exists in the universe, but he does not interfere in human affairs nor does he listen to our prayers and aspirations. We have free will without the 'will' of God.

A person who says they believe in a pantheistic God is different. Although pantheists believe that the universe was indeed intelligently created, and that a unifying force permeates everything, including all known (and unknown) aspects of the material world, it is not a supernatural entity. The term pantheism comes from the Greek words *pan* ("encompassing all things") and *theism* (a belief in God). A pantheistic worldview therefore encompasses the idea that everything is both interconnected and originates from the one substance – also known as monism. Humanity, nature, and all material creations are part of this unifying, numinous presence; it is within us and all around us. Some pantheists use the word divine (or God) when attesting to this presence, meaning that it is something presently beyond our understanding, and only sensed in our yearning for spiritual truth and meaning. That said, it is important to recognise that simply because some aspects of the universe are beyond our ability to comprehend does not mean they exist outside of universal physical law. There are no miraculous works for a favoured few, or a promised afterlife of endless pleasure.

A pantheist seeks rational and spiritual truth through understanding. Pantheism, therefore, is tolerant of all forms of personal spirituality that do not interfere in the lives of other individuals. It is not surprising that the innovators of modern pantheism, as it arose from the early Enlightenment of the 17th century, were deemed heretics by the Roman Catholic Church. Philosophers such as Giordano Bruno and Baruch Spinoza pursued the idea that a divine God was infinite and present in all things. They were subsequently ridiculed and rejected by Catholicism and Judaism, respectively. Bruno, an ex-monk, was even burned at the stake for crimes against the Church.

If taken to its end point, pantheism is indeed heretical to established notions of organised religion. For adherents of pantheism, there is no defined, personal relationship between man and the organising force which permeates the universe. This stands in direct opposition to traditional religious practice, where acts of prayer, worship, and sacrifice are made to a deity in the hope of material and/or spiritual salvation. The faithful are rewarded by a personal God who, in return, will protect his flock. Pantheism is different: the numinous also resides within each person, and this makes illumination and the search for meaning a personal journey of self-discovery. For the father of modern pantheism, Baruch Spinoza, it is the human mind that both projects and becomes the essence of this divine essence permeating the natural world. This vision of the divine has no place for doctrine or dogma. It also has no place for 'secret information.' Spinoza writes:

> As regards the human mind, I believe that it also is a part of nature; for I maintain that there exists in nature an infinite power of thinking, which, in so far as it is infinite, contains subjectively the whole of nature, and its thoughts proceed in the same manner as nature—that is, in the sphere of ideas. Further, I take the human mind to be identical with this said power, not in so far as it is infinite, and perceives the whole of nature, but in so far as it is finite, and perceives only the human body; in this manner, I maintain that the human mind is a part of an infinite understanding.[4]

Spinoza's conceptualisation of 'God' is also the God of Albert Einstein. Perhaps the greatest thinker of the modern era, Einstein believed that science and spirituality are not adversaries: they both seek to understand and therefore find meaning. He contended that rational individuals may feel imprisoned by the ultimate futility of personal desires and material possessions. They may seek "to experience the Universe as a single significant whole."[5] This need can be translated into either a spiritual quest or that of scientific enquiry. Einstein believed that the 'lower' aspects of religious belief, particularly those that advocate a personal God that demands prayer and sacrifice, has dominated religious practice and enquiry to the extent that the greater questions of meaning and existence are obscured:

> Thus one tries to secure the favour of these beings by carrying out actions and offering sacrifices which, according to the tradition handed down from generation to generation, propitiate them or make them well disposed toward a mortal. In this sense I am speaking of a religion of fear.[6]

Einstein even recontextualised religion and scientific thought as a unified "religious spirit of science."[7] He contends that the process of scientific enquiry can only really take place if the scientist is committed to a cosmic religious feeling:

> His religious feeling takes the form of a rapturous amazement at the harmony of natural law, which reveals an intelligence of such superiority that, compared with it, all the systematic thinking and acting of human beings is an utterly insignificant reflection.[8]

Both Spinoza and Einstein ultimately challenge us with the same question: What type of spirituality do we believe in? This is very different from asking what religion a person belongs to; in contrast, their challenge goes to the heart of things: Why do we believe what we believe? The God traditionally given us by all three Abrahamic faiths is one that demands personal attention, prayer, loyalty, and abeyance. This God is omnipotent and transcendent. This is not the deity of Spinoza or Einstein. Spinoza expressed the idea that the terms divine and nature are the same thing, and that humanity exists as a natural part of the universe. This perfect combination of the material and the numinous exists outside of any religious structure, in particular those religious beliefs that personify their deity as God incarnate. In proclaiming this, Spinoza cryptically references the square and the circle:

> All things, I repeat, are in God, and all things which come to pass, come to pass solely through the laws of the infinite nature of God ... The doctrines added by certain Churches, such as that God took upon Himself human nature, I have expressly said that I do not understand; in fact, to speak the truth, they seem to me no less absurd than would a statement, that a circle had taken upon itself the nature of a square.[9]

If, as Einstein suggests, time itself is relative, then what we achieve in this life impacts events across space and time. When Einstein was told of the death of a close friend, he replied, "Now he has departed from this strange world a little ahead of me. That signifies nothing. For us believing physicists, the distinction between the past, present and future is only a stubbornly persistent illusion."[10] Einstein also invokes Spinoza as his guiding light in his pursuit of a form of spirituality that does not compromise either his own values or the laws of science. He gives us a form of pantheism that expresses a belief in an organising presence that exists outside of our current understanding of material reality. In this worldview, there is absolutely no place for the incarnation of a singularly divine, supernatural saviour of humanity. For Einstein, the divine exists in the mysterious works of nature. "This firm belief, a belief bound up with deep feeling, in a superior mind that reveals itself in the world of experience, represents my conception of God. In common parlance this may be described as "Pantheistic" (Spinoza)."[11]

Figure 23: A mural depicting the luminaries of pantheism, by Levi Ponce. The painting, funded by the Paradise project, includes Albert Einstein, Alan Watts, Baruch Spinoza, Terence McKenna, Carl Jung, Carl Sagan, Emily Dickinson, and Nikola Tesla.

Sexed-Up Atheism?

Given the impact of the scientific revolution and the emerging symbiotic nature of science, technology, and human endeavour, it is not surprising that some people have openly rejected all forms of structured religion. During the early years of this century, a new adversary emerged to challenge the validity and relevance of organised religious belief - organised atheism. As previously noted, Richard Dawkins, in his book *The God Delusion*, takes direct aim at the institutions of all the great monotheistic religions. He argues that by definition, these faiths are 'theistic' in that they wholeheartedly embrace their God as creator and ruler over the universe. This God also influences the lives of his creations; he forgives, rewards, and punishes his people. Professor Dawkins then progressively argues that a belief in this type of God is both absurd and a destructive delusion that has continued to permeate Western civilisation. I concur with this argument. However, his views on deism and in particular, pantheism, need further discussion. For Dawkins, a pantheist is someone who sees the divine in the natural laws that govern the universe. Dawkins states that a belief in pantheism is really just "sexed up atheism" [12] and that deism is "watered down theism."[13]

Dawkins openly admits that a pantheistic vision of spirituality is problematic when discussing the relevance and role of religion. He therefore purposely directs his argument towards the man-made structures that have been created to disseminate religious dogma. It is important to note that he does not argue against a person's right to be a 'spiritual being.' He has even advocated the possibility for some sort of reconciliation between atheists and their Christian brothers. 'Atheists for Jesus' could celebrate the unselfish moral teachings of the human Jesus: "I think we owe Jesus the honour of separating his genuinely original and radical ethics from the supernatural nonsense that he inevitably espoused as a man of his time."[14]

Christopher Hitchens effectively partnered with Dawkins to inspire a vision of atheism that could stand in open opposition to structured religions. They did this in a way that could withstand any claims of high ground these faiths proclaimed regarding ethics, morality, and virtue. Sadly, Hitchens died in 2011. But he bequeathed us a legacy and testament of a person who was not afraid to stare bravely alone into the unknown frontier that death presents to us all.

Hitchens lived and died an atheist who completely discarded and to a great extent discredited the religious institutions of his time. In his book *God is Not Great,* we are given a vision of religion as a destructive man-made institution that feeds on people's fear of death and their need to exert control over things they have no control of. He confronts the leaders and believers of all three monotheistic faiths in equal measure without remorse or mercy. In doing this, Hitchens has taken an open and adversarial approach both in theme and action. His aim is not only to 'know the enemy,' but he is prepared to logically destroy the foundations on which they stand. From the book:

> I think we are entitled to three provisional conclusions. The first is that religion and the Churches are manufactured, and that this salient fact is too obvious to ignore. The second is that ethics and morality are quite independent of faith, and cannot be derived from it. The third is that religion is — because it claims a special divine exception for its practices and beliefs — not just amoral but immoral.[15]

He achieves his mission without destroying the innate dignity of humanity: we are flawed but on a communal journey of progressive evolution and self-discovery. Hitchens is prepared to challenge us to reveal (and re-evaluate) ideas and beliefs that for the most part are left unchallenged. Christopher Hitchens joyfully opened doors of perception into a new world devoid of any construction of a personal deity. Like Einstein, he also finds spiritual solace in the achievements of science and the humanist philosophy of Baruch Spinoza:

> Philosophy begins where religion ends, just as by analogy chemistry begins where alchemy runs out, and astronomy takes the place of astrology … Spinoza's definition of a God made manifest throughout the natural world comes very close to defining a religious God out of existence. … Projected forward into the mind of Einstein who answered a question from a rabbi by stating firmly that he believed only in "Spinoza's God" and not at all in a God "who concerns himself with the fate and actions of human beings."[16]

Up until recent times, pantheism has been sidelined to the fringes of any discussion on religious belief and spirituality. Many theists see it as a New Age form of paganism or even as an incarnation of Satanism. Many atheists see pantheists as confused non-believers who are unwilling to come fully to terms with their mortality. However, a belief in a pantheistic view of the spirituality may allow us (to some extent) to reverse the dilemma of Pascal's Wager. Blaise Pascal was a 17th century philosopher and scientist who is best remembered for his argument in favour of traditional religious belief. Put simply in terms of a yes/no decision, he argued that people should believe in a theistic God: "If God does not exist, a person loses nothing by believing in him; but if God does exist, belief in him can bring eternal life. Thus, one should wager that God exists.' [17]

However, most people who believe in God also belong to a structured religion. If they 'wager' to believe, they also pledge to be subservient to the rules of their religion. One life of obedience and servitude for an eternity in paradise! But what happens if we take away any notions of a God who demands prayer, sacrifice, and allegiance to a particular religion? In that case, we are each individually responsible for creating meaning in a world ruled by natural law. What happens after death is at yet unknown. A person loses nothing by seeking to make this life meaningful and enjoyable without having to 'bend the knee' to God. If eternal life exists, we have lost nothing by seeking a fulfilling existence here on Earth. Maybe we should "wager on the richness of life here and now."

Some religious apologists have also appropriated Einstein's vision of pantheism as a way of justifying belief in the personal deity of organised religion. Dawkins saw the danger in this and consciously refuted pantheistic ways of practising spirituality. From *The God Delusion*:

> the metaphorical or pantheistic God of the physicists is light years away from the interventionist, miracle-wreaking, thought-reading, sin-punishing, prayer-answering God of the Bible, or priests, mullahs and rabbis, and of ordinary language. Deliberately to confuse the two is, in my opinion, an act of intellectual high treason. [18]

A possible alternative to this dilemma would be to unequivocally place science and rationality at the heart of any definition and discussion concerning pantheistic-inspired spirituality. I believe that using the term 'scientific spirituality' would negate the religious connotations sometimes attached to pantheism. This could present an opportunity to present a new starting point in more precisely defining this form of spirituality.

As previously mentioned, spirituality can be defined and measured by a person's level of self-transcendence, that is, our genetically-sourced ability to observe the nature of the universe – and our place in it – outside the confines of our own personal needs and desires. In other words, to reach out beyond ourselves and search for some form of unifying numinous presence. A very brief definition of science as put forward by the Science Council states: "Science is the pursuit and application of knowledge and understanding of the natural and social world following a systematic methodology based on evidence." I contend that our inherited yearning to be spiritual beings is not only compatible with, but also interdependent with, our desire to find and understand universal scientifically proved truths. Spirituality is the predetermined driving force and science is the methodology by which our yearning may be satisfied. This is scientific spirituality.

Unlike pantheism, scientific spirituality takes the words *divine* and *God* out of its definition, and replaces them with the laws of science and the spirit of human endeavour and imagination. Here, everything in nature is connected, interdependent, and ultimately understandable. The practise of scientific spirituality also provides a gateway of understanding where we came from (our cosmology) and why we are here (our ontology and epistemology). I would therefore define scientific spirituality as the personally inspired pursuit of self-transcendence through the investigation of scientifically derived truths about the universe and our place within it.

Scientific spirituality also leaves a door open for non-religious theories of universe creation outside of the accidental chaos of evolutionary nihilism or even the more ordered meaning found in atheist existentialism. Atheism rejects any conceptualisation that our universe was purposely designed with a predetermined set of laws programmed to create super intelligence. This premise does, however, fit within the bounds of scientific spirituality. An important assumption behind scientific spirituality is that the universe has been designed by beings similar to us and that because of this there is indeed a predetermined plan to bring order out of chaos. This design includes both existing and emerging natural laws that are gradually being disclosed to us.

As previously discussed in Chapter Three, Edward Harrison argued that because our universe is ultimately understandable, and the constants of physics so finely tuned, it is reasonable to conclude that our universe was created by a superior intelligence existing in another physical universe similar to our own. I further contend that it is our yearning to be spiritual, and from this our desire to reach outwards, that is a primary catalyst pushing humanity towards transcendence. Our own descendants in the far future may even possess the knowledge to design and build universes. A future incarnation of humanity could fulfil their heritable yearning to be spiritual, with a quest for understanding, meaning, and enlightenment, that could culminate in finding the holy grail of cosmology: universe creation. Scientific spirituality embraces the idea that there could be a predetermined plan for humanity – or some augmented version of it – and that it may depend on our spiritual desire to reach beyond the confines of materiality. In other words, everything may start and end with Cosmological Natural Selection with Intelligence and Spirituality (CNSIS).

The Evolution of Spiritual Belief

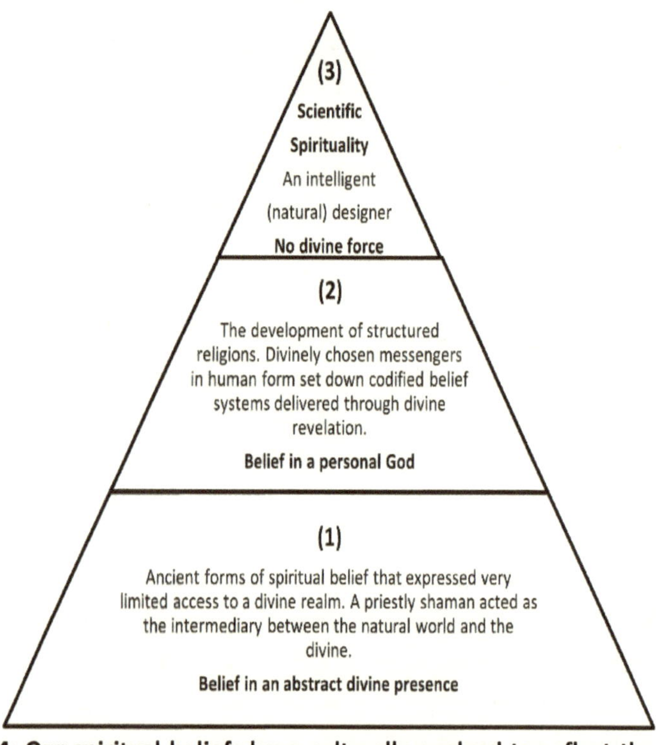

Figure 24: Our spiritual beliefs have culturally evolved to reflect the social and institutional constructs of their time. In a world now dominated by scientific truths and rapid technological advancement, our yearning to be spiritual beings is now focused on rational understanding instead of blind faith.

Key Points

- If, as the Greek philosophers concluded, a meaningful life is framed by a personal journey of enquiry and progressive illumination, the discoveries of science and the understanding of the cosmos must be deemed a good and meaningful contribution to the unfolding narrative and evolution of humanity.

- The empirical truths of natural law given to us by science have opened paths of enlightenment devoid of superstition and myth.

- An individual can now stand alone and experience the personal illumination of rational thought with a sense of spirituality that embraces all the mysteries of the universe.

- Our progressive evolution into more complex organisms, coupled with an innate drive to understand why and how things came to be, is at the heart of scientific spirituality.

- Science is not an adversary to this form of spirituality: it is a tool and methodology by which our evolutionary journey can be investigated and hopefully understood.

- It is the symbiotic nature of science and spirituality that pushes intelligent people to risk their lives when they embark on space missions in search of understanding the unknown, or dedicate their lives to teaching and medicine.

- Scientific spirituality rejects the supernatural for the natural, and substitutes the miraculous for the unknown wonders of the universe. If there is a unifying numinous presence, it exists within us and all around us.

Cosmological Natural Selection with Intelligence and Spirituality

*We must try to understand the beginning of
the universe on the basis of science. It may be
a task beyond our powers, but we should at
least make the attempt.*

Stephen Hawking, *The Universe in a Nutshell*

Introduction

Throughout the course of this work, I have endeavoured to discuss and form a framework of structured conclusions which, when layered over and appropriately interconnected, gives us a possible narrative about where we came from and why we are here. It does not inform us about what happens when we die. And until science gives us our first necro-naut who can visit the dead and return safely back, we will never know the answer to that question!

I have taken a compatibilist approach at three crucial points during this journey. First, I put forward the premise that free will and determinism can coexist – we are born burdened by our genes, but we have culturally (and to some extent physically) evolved to rebel against them. Second, I believe that intelligent design and natural selection not only coexist, but they are also interdependent. Finally, I propose that science and our desire to be spiritual beings are also interdependent and compatible. Listed below is a summary of the major ideas presented. If the reader rejects any one of these conclusions outright, then I believe that my entire premise – from the reader's perspective – should be deemed falsified.

1. The evidence of evolution through natural selection is overwhelming; who we are today is the result of millions of years of evolution. Genetic determination, up until recently, has been a primary catalyst for not only the human form but also for human endeavour and behaviour. Nature loads the gun of our behaviour, but nurture (our cultural environment) has the ability to not pull the trigger, or to aim the gun in a different direction.

2. Natural selection is the only natural process known to science that can generate *purpose*. It is only through non-random adaptive changes in organisms that genuine complex functionality can be created. Direct observation also suggests that there is some form of ordered design emerging from the cut-throat, chaotic, and amoral mechanisms of natural selection. For such a ruthless process to eventuate in the complexity we see in the modern human form does indeed point towards some type of inherent design and drive towards increasing complexity.

3.   Only superior intelligence has any hope in escaping the confines of natural selection and subservience to the natural world.   Darwinian natural selection in humans today has been somewhat neutralised, either through science, or by cultural intervention. Our culturally-evolved environments have also created and encouraged a degree of free agency. We are the sole species on this planet capable of genuine acts of free will.

4.   Our species is indeed unique, and destined at some stage to completely transcend the confines of Darwinian biological evolution and become masters of our own destiny, creating our own transcendent purpose. As a species, we are at the cusp of a second great leap forward that will result in a cognitive revolution inspired by the machine. This vision will be dependent on our ability to create a symbiotic collaboration between cultural evolution, our spiritual desires, and the intelligent designers of this new world.

5.   Evolution of superior (even super) forms of intelligence, (organic and/or artificial) may be part of a greater narrative, written before the birth of our universe. We may exist in a multiverse, where each universe could act like an independent organism with purposeful traits, and in which our universe is just one of many competing to reproduce itself to gain more representation in the multiverse.   One of the main driving evolutionary forces in universe creation and reproduction is intelligent life.

6.   Our universe has evolved according to a structured, pre-planned order of operations, ultimately designed to replicate itself – creating new cosmic habitats,

conducive to evolve intelligent beings with the ability to create other new universes. Those universes best suited to creating the trait of superior intelligence contribute to being the 'fittest' adaptions in universe creation. Humanity, or some augmented form of it in the future, may have the knowledge to design and build universes.

7. Although our religions may have initially been an expression of hope for something more than selfish survival, they also nurtured a deluded meme that a personal God ruled over everyone and everything. Religious beliefs in a supernatural deity and the blind obedience given to a God-chosen representative – usually a dominant male – has made religion a culturally-inspired, totalitarian weapon for the replication of those selfish genes best suited to ruthlessly rule over others.

8. Each of us is genetically tuned to personally yearn for something beyond tangible materiality. We are by nature spiritual beings. Although this trait for spirituality has been culturally warped and abused, it still pushes each of us to search for meaning. If we disregard structured religion, we stand alone – empowered to create our own meaning. Our imagination, sense of truth and beauty, compassion, and spiritual desire to seek understanding of the world around us may play an even more important role in creating a meaningful life.

9. Our inherited yearning to be spiritual beings is interdependent with our desire to find and understand universal scientifically-proved truths. Spirituality is the predetermined driving force and science is the methodology by which our yearning may be satisfied. This is scientific spirituality. Everything in nature is connected, interdependent, and ultimately

understandable. The practise of scientific spirituality also provides a gateway of understanding where we came from, and why we are here.

10. Our yearning to be spiritual is the primary catalyst pushing humanity towards transcendence. Science and superior/super intelligence create the ability to transcend this world, but it is our spiritual yearning that drives us to reach for it. Our own descendants in the far future may design and build new universes. Scientific spirituality also embraces the idea that there could be a predetermined plan for humanity – or some augmented version of it – and that it may depend on our desire to reach beyond the confines of materiality, to Cosmological Natural Selection with Intelligence and Spirituality (CNSIS).

Where Did We Come From?

If we take all possible eventualities into consideration about the creation of our universe, it comes down to three radically different alternatives. First, the universe has been created supernaturally by a divine entity (the theistic principle). Second, everything exists as the result of natural occurrences devoid of any divine, creative, or naturally-inspired organising force (the atheistic principle). Finally, our universe exists because it was created by intelligent beings – most probably, beings similar to us (the anthropic principle). I provisionally believe that the last alternative, given the circumstances in which humanity now finds itself, is the most likely scenario. I would even extend this alternative to include two possible creation scenarios:

A. *We exist in a finely-tuned universe, designed by intelligent and spiritual beings similar to us, as part of a greater mother* *multiverse, where offspring universes that are most hospitable to intelligent/spiritual organisms are naturally selected to design and build new universes.*

or

B. *We exist in a finely-tuned universe, designed by intelligent and spiritual beings similar to us, but who exist in our far-off future, and who have encoded and communicated a set of natural laws back to the point of universe creation that are most hospitable to the cosmic evolution of worlds conducive for the biological evolution of intelligent/spiritual organisms.*

Scenario **A** envisions that our universe is one of many universes, and that the forces of natural selection begin at the point of universe creation – much like that envisioned by Smolin and extended on by Price, but also including our spiritual yearning as a prime catalyst. **Scenario B** expresses the idea that there is only one universe, and that natural selection begins with stellar evolution (the creation of galaxies, stars, and planets). This scenario is dependent on our ability to engage in some form of time travel that will enable intelligent seeding of our universe at the point of creation. Both scenarios, however, are dependent on an intelligently designed universe (or universes), albeit at different points in universe creation. Both are dependent on an initial spark of creation that has been purposely encoded in natural laws. Both concur with the anthropic principle, and both exist within the domain of what can be termed Cosmological Natural Selection with Intelligence and Spirituality.

As previously introduced, and simply stated, the anthropic principle infers that the universe is the way it is because we exist. However, by definition this premise is circular

– put the other way around, we exist because the universe is the way it is. The only way to validate an idea that is circular is to show that the logic behind the premise is itself valid and culminates in a conclusion that initiates the first cause of the premise. In other words, the first and last parts of the theory are wholly dependent and interdependent. We must, therefore, create a theory of universe creation that directly links intelligence (and I would argue, spirituality) within the domain of evolutionary natural selection. We need also to provide proof that can be empirically evidenced. Based primarily on the pioneering work of Edward Harrison, the following discussion will deliberately focus on an anthropic principle emphasising Cosmological Natural Selection with Intelligence and Spirituality. Again, I will use the term *intelligent design* to describe interventions in the natural world by intelligent beings, not God.

A universe must have a creation point. I contend that this cosmogenesis has been intentionally and intelligently designed to kick-start natural processes under the chaotic guidance of natural selection – first, cosmological (stellar) selection, and then, biological selection. This design spark initiated the fundamental laws and constants of physics so that in some universes (or parts of universes) intelligent beings would someday come into existence. A complicated and ironic point of reference here is that the pinnacle achievement of intelligent transcendence is to initiate a bottom-up process – evolution by natural selection – as the mechanism by which complexity can come into existence.

One of the strongest arguments for intelligent intervention in the initial creation of our universe is that the basic constants – the speed of light (c), energy/charge (e), mass (m), Planck's constant (h), and the gravitational constant (G) – are so finely tuned. Stellar evolution – the evolution of galaxies, stars, planets, etc. by natural means – could not have occurred unless the fundamental constants were perfectly aligned.[1] Dramatic changes in the composition of the universe can occur due to exceedingly small changes in the physical constants. For example, 'when (G) is reduced, stars cease to be luminous; when increased, they burn too quickly and their luminous lifetimes are too short for biological evolution.'[2] Conversely, a small increase in (e) 'creates a universe without elements other than hydrogen, elements that are vital to life.'[3]

Harrison also uses the term *natural selection* for the subsequent events in star/planet formation that culminate in an environment conducive to organic habitation. Even though the initial process was programmed for self-directed selection, naturally-induced variations are essential in creating environments 'fit for life.'[4] It could also be argued that there are two interconnected domains of natural selection at play. Fitness, from the perspective of biological evolution, concerns itself with an organism's adaptability to an environment and reproductive fertility. In biological evolution, there are 'two units of selection': vehicles (the living organism that survives long enough to pass on genetic information) and replicators (the actual genes that have been delivered by the vehicle). In Harrison's cosmology, the two components are "the habitability of the universe and the intelligence of the inhabitants."[5]

Harrison's work concurs with Smolin's theory that black holes may be the vehicle for universe creation, but both he and Michael Price also emphasise the importance of intelligence as the prime replicative device. However, as expressed in previous chapters, I would argue that there is another linking mechanism: our inherited yearning to be spiritual beings. I also place importance on our species' ability to first transcend the confines of Darwinian natural selection, and to finally reach for transcendence over the material world. This can only be envisioned by the working of creative imagination, which, although it exists within the domain of natural law, can see beyond the limits of what is presently known, and can embark freely into the unknown – culminating in the ultimate act: universe creation.

Edward Harrison does not mention the possibility for time travel in his theory of universe creation, but he does contend that our own descendants in the far future, may possess the ability to build universes. If time travel – even for quantum particles – is possible, then it may not be unreasonable to conclude that it may be own super intelligent and spiritually inspired designers who will design and deliver the spark of creation to this universe – thus becoming our own 'mother' universe. At some future point, intelligent life may be able to send an instruction code of defined natural laws back in time, and initiate a predetermined playing field through which the ruthless and amoral workings of natural selection can flourish. This in turn, brings about the possibility for the evolution of intelligent life. Only intelligent life can experience spirituality and imagine new ways of being. Only intelligent, spiritual and culturally enlightened beings are capable of genuine acts of free will and therefore possess the ability to transcend the confines of determinism.

The Process

Within the domain of the anthropic principle, we now have two alternative ways for universes to evolve: Through cosmological selection, beginning via a multiverse, where each universe acts like an independent organism with purposeful traits, competing to reproduce itself to gain more representation in the multiverse. Those universes most conducive to creating organic life, and from this, intelligent life, are naturally selected to reproduce. Alternatively, if the spark of creation comes from our own distant future, then cosmological selection begins at the point of galaxy and star creation – stellar evolution. In both alternatives, Intelligent design and evolution through natural selection work in tandem, from initial cosmogenesis to the creation of intelligent life.

Where we came from and why we are here is not exclusively due to either evolution or intelligent design, but rather to an interplay between both. Our universe was intelligently created – through purposeful cosmogenesis. Predetermined natural laws created environments capable of independent development, where evolution through natural selection could produce complexity through bottom-up processes over an exceedingly long period of time: First, through stellar evolution, culminating in the formation of worlds conducive for housing organic life – which in turn initiates biological evolution. Eventually, some life forms reach beyond survival and strive to be intelligent and spiritual beings, who are guided by (but not ruled by) their progressive cultural creations. Once an organism is capable of transcending the confines of natural selection completely, it can design complex and even super intelligences – either organic or artificial –

through top-down processes ultimately capable of universe creation.

The following diagrams detail the three separate systems of evolution: Stellar, biological, and cultural. Each system operates under the confines of natural selection. This means that each system must initiate a bottom-up process of non-random adaption that results in survival of those organisms (physical and biological) best suited to their environment. For this to occur, each system must also posses appropriate vehicles, replicators, and, as I have argued, some form of ultimate purpose.

Intelligent Design and Evolution (CNSIS)

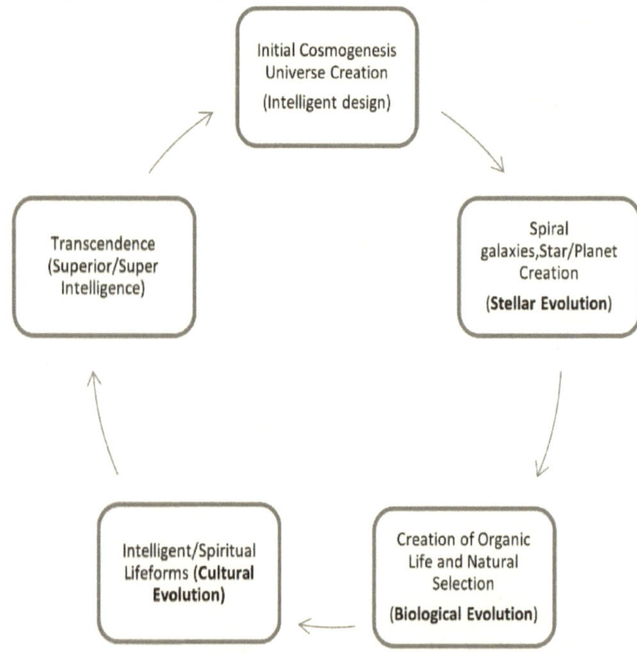

Figure 25: Intelligent design provides the initial spark of creation. Evolution (stellar, biological, and cultural) is the mechanism working towards the creation of intelligent and even super intelligent life.

Three forms of evolution By Natural selection

Evolution	Vehicle	Replicator	Ultimate purpose
(1) Stellar Spiral Galaxy	Stars	Planetary systems	Planets habitable for organic life
(2) Biological Habitable planet organic life	Living organisms	Genes - cellular reproduction	Intelligent life
(3) Cultural Intelligent life	Communities/ social systems, spirituality	Cultural memes; moral/ethical codes	Transcendence from natural selection/materiality

Figure 26: Cecilia Helena Payne was the first person to identify that stars are mostly composed of hydrogen and helium. One of the pioneers of the study of stellar evolution, Payne observed over three million stars. In a profession overwhelmingly dominated by men, she became a role model for other women interested in studying astrophysics.

Other Evidence for Design

My conceptualisation of CNSIS infers an interplay between intentional design and the workings of natural selection. Our universe was initiated by an intelligent designer who made sure that the basic tenets for the creation of intelligent life in the far-off future would be possible but not assured. As previously mentioned, this 'top-down' ignition then allowed the universe to evolve according to natural law, which includes the ruthless long reach of natural selection. I contend that the bottom-up complexity achieved through the rigors of evolution by natural selection is itself, paradoxically, proof that a predetermined order of operations exists in the universe. Our designer made the dice by which natural law is confined, and evolution through natural selection is the game played.

In 2002, mathematician Stephen Wolfram demonstrated that a set of simple rules, "just one or two lines long," can generate immense complexity. He further speculates that "the code required to generate the entire universe may also be just a few lines long."[6] As a monist, I believe that everything in the universe is governed by the laws of physics. Therefore, as expressed in our second (and my preferred) scenario, any laws present at the start of our universe will be encoded at some point in our distant future and sent back in time as a blueprint for igniting the forces of evolutionary natural selection. Circular reasoning indeed, but logically possible. Every new breakthrough in our understanding of physics takes us one step closer to this goal.

Alternatively, in the multiverse scenario, intelligent design initiates the ability to create new universes which will naturally select those universes capable of sustaining intelligent life. Scenario B (time travel) goes against special relativity and pushes the boundaries of general relativity, but still may be possible. Conversely, the multiverse scenario encounters the problem of who or what created the first 'Mother' universe. Apart from saying 'God did it,' or that everything exists from the accidental chaos of the Big Bang, I believe CNSIS represents our best logically-inspired theory of explaining how things came to be.

Interestingly, Edward Harrison also posited that each human civilisation throughout history created their own vision of what it meant to exist in a 'Universe.' These creations reflected knowledge as it was then understood, and because it provided meaning, it was deemed true. Harrison also used the term Universe (note the capitalisation) to describe the unknown – which represented, for these early civilisations, the unknowable – aspects of the real Universe that existed beyond reach. From his book *Masks of the Universe*: "Always only a few things remain to be discovered. We pity the universes of our ancestors and forget that our ancestors will pity us for the same reason."[7]

For those people who continue to believe in a theistic principle of universe creation, appendixes at the end of this book summarise the major Western creation stories. I invite the reader to weigh these against my 'exotic' scenarios. I have not discussed other religious traditions and their stories of creation because, firstly, I have limited understanding of them; also, because they ultimately believe in some form of supernaturalism and/or dualism, I cannot find a platform from which to investigate them.

In 1983, physicist John Wheeler posed the question, "Is the machinery of the universe so set up from the beginning, that it is guaranteed to produce intelligent life at some long-distanced point in its history-to-be?"[8] Apparently, Wheeler was motivated by an intuition "that we do have a special relation to the universe – that intelligent, conscious beings are in some sense necessary to the universes existence."[9] This proposition becomes starkly apparent when we consider the importance of observation and observer perspective in both relativity and in quantum mechanics. It could even be argued that the universe and the observer "exist as a pair."[10] In other words, we live in a participatory Universe and we are the main participant.

Within the microscopic domain of the quantum subatomic world of matter, according to the most popular theory – the so-called Copenhagen interpretation – subatomic (or quantum) "particles do not exist in one state or another, but in all of their possible states at once."[11] It is only when someone observes a quantum particle that the particle is forced to choose a probability of motion, and it is this motion that the observer sees. In other words, a quantum particle does not have a location until it is observed. And even then, by observing one particle, its correlating twin particle is also affected, "no matter how far apart they are in a process known as Quantum entanglement."[12] This event happens instantaneously, even over large distances, and therefore occurs faster than the speed of light.

In special relativity, with respect to larger cosmic forces, observer perspective is the frame of reference by which events/objects can be measured. Einstein used the example of an extremely fast train which is struck simultaneously at the front and back by lightning, as observed by a stationary bystander who sees the event when the train is midway through its journey past him. However, another person on top of the train as it travels would see things differently. She would observe the lightning bolt at the front of the carriage first and the second one sometime later. Why? Because as the train moves forward, the light from the lightning bolt which hits the rear of the train has to travel a further distance to reach the observer on top of the train. Therefore, from the perspective of the first (stationary) observer, time for the second observer, who is moving away, is passing more slowly. This is known as *time dilation* – the faster an object moves, the more slowly time moves. In other words, the faster an item (*m* or *mass*) travels, the slower time flows. Hypothetically, if a person can travel fast enough to reach the speed of light, then time would stand still.

Einstein also used an example of viewing a clock while travelling away from it at the speed of light. Because we see the same beam of light constantly, the clock's hands would not move – time would stand still. Therefore, there is no constant or measure of time that is universal. Instead, time stretches and contracts, varying with velocity. As an object accelerates towards the speed of light, its length contracts and its mass increases. This is known as *length contraction*. The twin occurrences of time dilation and length contraction ensure that the speed of light is constant for all observers. Time, and the space it travels through, forms its own dimension: spacetime. In special relativity, it is theoretically impossible to exceed the speed of light. At or near the speed of light an object's mass increases dramatically and requires a near infinite amount of energy. Einstein's ground-breaking connection about the relationship between matter (m), energy (E), and the speed of light (c), was summarised in the equation, $E = mc^2$.

The general theory of relativity expanded on Einstein's previous work by including the effects of gravity and acceleration on spacetime. Objects with large amounts of mass warp space/time, causing it to become curved. This in turn causes objects to experience gravitational attraction to each other. For example, objects in the fourth dimension of spacetime attempt to move in straight lines, but because spacetime is curved, their trajectories appear bent.

Objects with large amounts of mass, such as stars and planets, also affect the relative measurement of time. The closer a person is to an extremely large object, the more slowly time passes. It could therefore also be possible for someone to comparatively travel forward in time by being close to a very large object with immense amounts of gravitational attraction. They would experience time at a slower rate than their compatriot on Earth. Upon return, they would now be at some point in the future. Space and time, and their relationship to an object's mass, do not exist separately; they are unified. Every moment of our lives has a particular reference in space and time, and we have to manipulate both to travel in time. Interestingly, the immense amount of gravity experienced near the vicinity of black holes, where an extreme amount of mass is contained in a small amount of space, causes time to slow to a halt at the entrance of the dying star – even light cannot escape its confines.

The last book authored by Stephen Hawking, *Brief Answers to the Big Questions*, was posthumously released in 2018. In it, he spends an entire chapter on the possibility of time travel – in particular, the supposedly impossible dream of travelling back in time. Hawking convincingly argues that Einstein's general theory of relativity does not rule out this possibility, although the amount of energy needed to exceed the speed of light – the requirement to travel back in time – "would take an infinite amount of power to accelerate past the speed of light."[13] Hawking argues that wormholes which have the ability to "warp space-time in the opposite way to that in which normal matter warps it" could prove to be a gateway for travelling into the past. "So it might seem that as we advance in science and technology we might be able to construct a wormhole or warp space-time in some other way so as to be able to travel into our past."[14]

Hawking concedes that there are problems with travelling into past history. He is acutely aware of the contradictions, paradoxes, and logical obstacles of this theory. An obvious example of this dilemma is the possibility that someone could go back in time and kill their younger self and therefore defy the laws of nature and logic. Hawking, in answer to these problems, argues that time travel into the past may be limited to micro 'virtual particles' that have no tangible history in their present state. He refers to this limited possibility as "the chronology protection conjecture."[15] This infers that history (past and present) remains fixed and supports "a consistent histories hypothesis" regarding time travel.[16]

I do not propose that any of the alternatives presented here represent a definitive answer to how our universe was created. Instead, I have heeded Hawking's advice, and attempted to create a theoretical framework that can be objectively discussed, rejected, or even laughed at. But at least I have made 'the attempt,' albeit through referencing the inspiring work of some brilliant scientists. One thing that I will again confirm, and take ownership of, is my near certain opinion that a personal God **did not** make everything or anything. It is up to us as to where we journey next – both internally and as a species.

Figure 27: Stephen Hawking (1942-2018) opened new doors of perception about how we view ourselves and our place in the universe. His final work even speculated the possibility of time travel into the past. Hawking is best known for his pioneering research on black holes, and for his ability to communicate complex scientific ideas in a way that could be understood by non-scientists.

Transcendence

As a species, we are standing at the precipice of a new way of being which bears little resemblance to past cultural or technological creations. The monuments of our ancestors that still stand today are mostly religious in nature. Anyone who visits the oldest and longest lasting civilisation, that of ancient Egypt, will be confronted with religious icons in tribute to dead Kings and Gods. Even today, most Western cultures are trapped in a paradigm that starts with the premise that an all-knowing, all-loving God rules over his creations. Many of our philosophies also begin with this idea, and only then do they construct theories dependent on the intrinsic dualism implied by their beliefs in supernaturalism. Our heads are not only shrouded in the blurry fog of faith, they are also steadfastly pointed backwards to a world that no longer exists.

If super intelligence is the trait that rules the world to come, then I argue that the best evolved traits of humanity – our intelligence, compassion, imagination, and sense of spirituality – should be an important part of this future. This will only be possible if we can create a moral framework based on individual and communal well-being that is personally empowered, culturally inspired, and scientifically accountable. I believe that this could be the biggest paradigm shift in the history of our species. In the past, our actions were accountable to a God who would reward (or punish) us in this life or the next. Instead, we could embrace a morality dependant on empirically confirmable (or falsifiable), scientifically accountable, truths. But here's the rub: science can tell us if the morals we use contribute to well-being, but it does not tell us what specific morals we should value individually or communally. It will only inform us if the values and morals we choose contribute to well-being.

Artificial intelligence will be part of this new reality. We do not know if we will dominate it, or it us; hopefully, the world we create will be compatible, not confrontational. If we want to include artificial intelligence in creating an inclusive moral framework, do we get it to memorise the most ancient book of monotheism, the Old Testament, and use it as the primary scaffold of creating those things we will jointly value? Given the contents of that book, and the lessons not learned by humanity, our artificially created comrades may just decide to destroy us all and start again – the God of the Bible has form in this area!

In the 1960s, scientist James Lovelock proposed the idea that the earth itself was a living organism; he called this proposition *the Gaia hypothesis*. Lovelock argued that because the earth was a self-regulating organism, this represented evidence that it is, in a broad sense, a 'living' system, operating autonomously through a series of interacting bio-geo-chemical cycles from a planetary perspective. However, this does not infer that the earth is either sentient or conscious. Without resorting to notions of supernaturalism, Lovelock has given us another pathway to expressing not only our personal spiritual desires but also the possibility that as part of a greater living organism, our sense of spirituality may be entwined within a communal evolutionary journey towards the preservation of life. As Lovelock declares:

> Gaia theory does not contradict Darwinism, rather it extends it to include evolutionary biology and evolutionary geology as a single science. In Gaia theory, organisms change their material environment as well as adapt to it. Selection favours the improvers, and the expansion of favourable traits extends local improvement and can make it global. Inevitably there will be extinctions and losers, winners may gain in the short term, but the only long-term beneficiary is life itself.[17]

If our material world is indeed part of a greater living and evolving organism, then artificial intelligence – because it is a material creation – is also part of this greater living entity. We participate directly in biological evolution and also as a component part of stellar evolution. Lovelock concludes that we may be destined to merge with AI as part of an evolutionary drive within both humanity and the planet: "I think like all organisms on Earth our species has a limited lifespan. If we can somehow merge with our electronic creations in a larger scale endosymbiosis, it may provide a better step in the evolution of humanity and Gaia."[18] If taken to its end point, our destiny may not only be to be able to design and create new universes, but also to travel within them. Artificially evolved versions of ourselves may indeed transcend the confines of materiality and break away from our parent organism, the earth itself, and explore even larger self-regulating systems.

Research by Michael Burton, from the University of New South Wales, contends that Darwinian forms of natural selection exist on four different levels within the universe. These include, "in order of increasing magnitude, the cell, the forest, the Earth (Gaia) and the Galaxy."[19] Each of these act as self-regulating ecosystems in which the forces of natural selection ensure survival of those organisms best fit for their environment. Interestingly, Burton confirms the specific role of spiral galaxies as the primary vehicle for stellar evolution. "Our Galaxy, and spiral galaxies in general, must be the first ecosystems to form within the Universe, within the first 3 billion years of its existence. They are thus the simplest such systems to exist within it, with their operation determined entirely by physical processes."[20]

Religiosity Without God – Scientific Spirituality

If our religious traditions are to survive the new world to come, they will need to come to terms with a reality based on scientific truth. At best, they could act as a conduit encouraging harmony and spiritual development. For example, in his encyclical letter of 2015, Pope Francis made a formal plea to all people regarding the future of the planet and humanity. In this address, the Pope embraced the natural world as a feminine creation and even implied that damage to the environment has been done by inherently male 'lords and masters' in their abandonment of Mother Earth.

> This sister now cries out to us because of the harms we have inflicted on her by our irresponsible use and abuse of the goods with which God has endowed her. We have come to see ourselves as her lords and masters, entitled to plunder her at will. The violence present in our hearts, wounded by sin, is also reflected in the symptoms of sickness evident in the soil, in the water, in the air and in all forms of life. This is why the earth herself, burdened and laid waste, is among the most abandoned and maltreated of our poor; she 'groans in travail'. We have forgotten that we ourselves are dust of the earth; our very bodies are made up of her elements, we breathe her air and we receive life and refreshment from her waters.[21]

This inspiring letter reinforces not only the Church's traditional belief that the world and the earth itself are inherently good but also emphasises the responsibility we all have, to each other and to the planet. If we remove the word 'God' from this message, it becomes an expression of scientific spirituality; it is not the dissemination of supernatural dogma. If our religious traditions are to survive in some tangible form, their reliance on mythical constructions, and supernaturalism in all its forms, not only need to be discarded but also openly recognised for what they are: tribally-inspired myths. As renowned historian Elizabeth Vandiver concludes:

> Myths are ostensibly 'true', that is, they present themselves as giving an accurate narrative of 'what really happened'. A culture rarely recognises its own mythology as mythology. Judged from within a culture, myths are true accounts of the way things really are.[22]

It is most certainly time we stand up, walk outside our religiously-embedded comfort zone, and see our myths for what they really are.

Maybe Richard Dawkins' call for a new tribal community consisting of 'atheists for Jesus' could also be part of the world to come. If we strip away the supernatural aspects of the Christ myth and disregard the institutional constructions that arose around him, all that remains is a story of someone born in poverty and raised in obscurity, who dedicated his life to being a voice for the oppressed and abused. His works and words give people the hope of something better, something more than mere servitude and survival. Jesus rebelled against organised religious sects and political structures and gave hope to the hopeless that the struggles of life are not in vain. This is the tangible legacy given to us by the Jesus of history. Seen through this perspective, the story of the human Jesus is part of a greater narrative that puts the individual at the centre of all things.

The man-made structures of religion, government, monarchical power, institutional abuse, and other mechanisms which ultimately have the power to either enhance or diminish human dignity and agency are again at a critical turning point in Western society. Our achievements in science, art, literature, education, medicine, architecture, social cohesion, and secular democracy can only stand if each of us recognises that our personal struggle is part of something greater. Spinoza and Einstein have given us a vision of spirituality that may withstand the onslaught of all forms of religious fundamentalism; they encourage us to seek our own unique pathway through space, time, and materiality — a uniquely Western-inspired vision of scientific spirituality. Transcendence of our inner self from the genetic chains of our ancient past. Transcendence from the material confines and brutality of natural selection. And maybe even transcendence of the natural world for new realities.

An Infinite universe

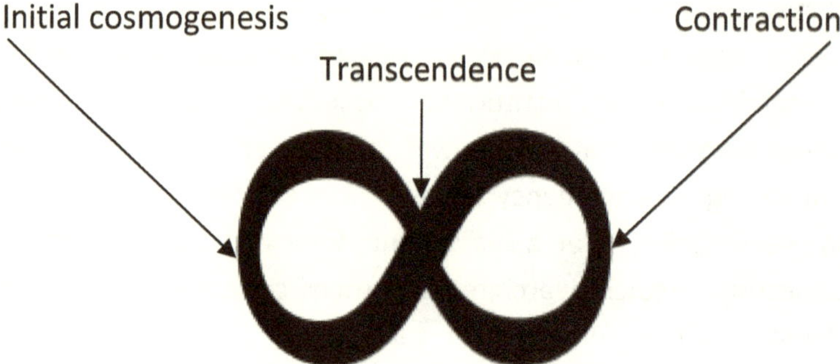

Figure 28: The symbol for infinity was originally envisioned by English mathematician, John Wallis in 1655. He uncannily also expressed the idea of a universe expanding and contracting over time. Our point of transcendence occurs at the intersection of space/time where an intelligent spark of creation can be communicated back in time to the point of cosmogenesis.

Appendix A: Ancient Western Creation Myths

Egypt

Ancient Egypt is the longest surviving ongoing culture in history. The main reason for this is probably the relative isolation of the Egyptian people from other cultural influences. The fertile plain of the River Nile provided an environment that could sustain a growing and complex society, while the adjacent desert ensured the ancient Egyptians were isolated within the confines of the upper and lower Nile regions. The first dynasty of consolidated tribes under Menes occurred around 3050 BCE, when upper and lower Egypt were joined into one united kingdom, with the city of Memphis as its capital.

From the beginning, the Egyptians saw the divine in nature, in particular, in the sun, the stars, and the River Nile. Most of the information historians have about the very early religious beliefs of the ancient Egyptians is found in the texts which appear on the earliest pyramids at Saqqara near Cairo. These date back to around 2800-2700 BCE. Famed Egyptologist James Breasted describes the hieroglyphs found in the pyramid texts as the first expression of abstract reasoning by humankind in writing. Here, out of chaos, the sun God Atum appeared. This creator God brought Ma'at , the concept of truth and order, into the universe. Ma'at acted against inherent disorder and potential chaos from darker spiritual forces. It was from the chaotic waters of Nun that Atum stood on the Benben and calmed the chaos through Ma'at. The sun God begat Shu the air God and Tefnut his wife. From these came the earth God Geb and the sky Goddess Nut. They created the earthly domain that gave birth to the first four siblings, brothers Osiris and Seth and sisters Isis and Nephthys.

Over time, the sun God Atum evolved into the deity Ra or Re, who later became the supreme Egyptian deity Amun-Re. Amun-Re became the protector of the pharaoh as well as the protector and guide of all people in their journey to the afterlife.

While they mostly adhered to the general explanations given in the pyramid texts, each major city in Egypt had its own variation of the creation story. In the creation myth of Elephantine, the supreme God was called Khnum and he alone created all living things. Another variation is the creation myth of Hermopolis, in which four pairs of male and female Gods emerged from the primordial wasteland to create a cosmic explosion that created all things. This group of eight creator Gods was known as the Ogdoad. Finally, the Memphis creation myth holds that the God Ptah used words to bring the universe into existence. Other Gods and humans were created by Ptah simply by saying their name.

Stories and mythical accounts of the Gods were important to the ancient Egyptians because they helped consolidate religious belief as a part of everyday life. These mythical constructions also acted as a moral compass that guided the Egyptians not only through a productive and ordered life, but also on their journey to the afterlife. The most significant and enduring myth passed down by the ancient Egyptians is the story of Osiris. Osiris, his consort and sister Isis, his sister Nephthys, and his brother Seth are the first deities to live within the earthly realm of existence. Osiris represents the light or good essence, which is opposed by the forces of darkness that dominate Seth. Osiris is warned by Atum of Seth's potential treachery but this does not prevent Seth from killing Osiris. As the myth tells us, 'His brother Seth felled him to earth.'

Osiris's body was dismembered and scattered across the earth. A devastated Isis searched for the body parts. She was eventually able to reconstruct his whole body except for his penis, which had been eaten by a fish. Isis fashioned a rod as a penis and became pregnant by the dead Osiris. Their son Horus was born and became the avenger of his father's murder by Seth. During the battle between Horus and

Seth, Horus lost his eye. It was salvaged by a new God, Thoth, the God of wisdom, who spat on the eye to heal it. The other Gods, upon hearing the desperate cries of Isis and her sister, chose to resurrect Osiris. They demanded that Osiris 'raise thee up.' Osiris joined with Thoth and the nine Gods of the divine Ennead. Osiris, as a resurrected God, had no place in the earthly realm and became the guardian of the underworld.

The Osiris myth, as described in the pyramid texts nearly 5,000 years ago, is the first written description of the institution and obligations of family. According to Breasted, it also reflects a degree of moral consciousness, as seen in the sacrifice of Horus as well as the fidelity shown by Isis to Osiris and Horus. The Osiris myth has other, more contemporary versions but, in essence, has remained intact and faithful to the pyramid texts. The resurrected and triumphant Osiris was later celebrated in the Sed Festival, in which the pharaoh assumes the persona of Osiris. Likewise, the eye of Horus, lost in the battle with Seth, became one of the most revered symbols in the Egyptian religion, along with the sacred scarab and the ankh. Horus changed in both intent and appearance over time. He was originally depicted as a warrior with the head of a falcon, or simply as a falcon, symbolising his dominance of the air. His name translates as 'he who is above.' The eye of Horus became popular as a protective amulet, representing the God's bravery and strength. By the later stages of the Egyptian kingdom, Horus is represented as a boy or an infant on Isis's lap, symbolising the combined strength of mother and child. Isis herself was deified not only in Egyptian culture, but was also directly and indirectly associated with the divine by Greek and Roman societies.

Mesopotamia

In modern-day Iraq was once a civilisation that rivalled that of the ancient Egyptians. A diverse mix of peoples and cultures over time became known as the Mesopotamians. Situated in the fertile plain between the Tigris and Euphrates rivers is the Tigris-Euphrates Valley. Its first identifiable settlers were Neolithic tribes known as the Ubaidians, who lived there as early as 4000 BCE. There is also evidence that the Sumerians, a more sophisticated civilisation, originally entered Mesopotamia in around 3100 BCE. They introduced cuneiform writing and the concept of organised rule through both kingship and a basic bureaucracy. It was not until the rule of Sargon the Great in around 2330 BCE that the disparate tribes of Mesopotamia became a unified kingdom under the rule of the Akkadian people, descendants of Semitic Amorites. Sumerian supremacy over the region returned in 2111 BCE under the rule of Ur-Nammu. The constant movement of peoples served to fuel an environment of conflict between the different tribes and cultures that wished to make this region their home.

There is evidence for two separate myths of creation that can be attributed to ancient Mesopotamia. The first originates from the Akkadian era (around 2000–1500 BCE). The earth God Ea, or Enki, sends seven sages to Earth to teach humanity important aspects of living in a civilised society. One of these sages, Adapa, is a priest. He is summoned by Ea, who informs him that the supreme God Anu has requested that Adapa appear before him. Adapa, on advice from Ea, refuses to partake of the sacred bread of life, believing it is the bread of death, and therefore misses an opportunity for eternal life. Humanity is now doomed to a mortal existence of birth, life, and death.

The second creation myth, which is evidenced in much more detail, comes from the Babylonian era. It is in the form of a poem composed in around 1100 BCE, but based on an older story created by the Semitic forefathers of the Akkadians. The poem, *Enuma Elish*, describes a beginning that is dense with a sense of boundless nothingness. Even "the Gods were nameless, natureless, futureless."

Three creation Gods emerged from this wasteland: Apsu (sweet waters), his wife Tiamat (salty sea) and Mummu (an opposite binding force). These intangible creation Gods made more Gods, including Lahmu and Lahamu (salt and water), Anu (heaven), and Ea (earth). The process of creation and the Gods' attempts to make order out of chaos were a constant struggle and they fought among themselves. It was only when the God Marduk, the perfect son of Ea, emerged and disposed of Tiamat that order was achieved. Marduk split Tiamat's body into two, and created a new earthly realm and a divine heavenly realm in the sky. Marduk became the king of the Gods and he alone could wear the Tablet of Destinies on his chest. Marduk created humanity from the blood of the slain monster Qingu. The intended purpose of humanity was to serve the Gods without question in all things. However, Marduk used divine blood mixed with earthly mud to create the human form, and a tangible link between humanity and the divine was established.

This creation story is one of constant struggle, chaos, and violence. Order is not easily gained. This theme may reflect the chaotic evolution of Mesopotamian culture as well as the unpredictable natural environment, influenced by the irregular flooding of the twin rivers. The Gods themselves are organised in a hierarchy. Anu, the supreme heavenly God, rules over the Anunnaki, his offspring, who, in turn, rule over lesser Gods. Ea is the creator God of earth, and the city of Babylon is placed at the centre of this new earth. The great God Marduk rules over heaven and earth. However, both the Gods and humanity were subject to laws that ensured the survival of order and the submission of chaos. Humanity and divinity are linked and under the same rule.

The power of the feminine and of fertility was important in ancient Mesopotamia, as personified by the Goddess Inanna from the Sumerian era. In the late Babylonian Empire, she was known as Ishtar. Inanna became a powerful Goddess of heaven and earth and was called the 'lady of myriad offices.' She ruled over sex, sexuality, and fertility (both in nature and humanity) and had a special interest in the fortunes of the people in her earthly realm. Although powerful and benevolent, Inanna was also passionate and tended to be chaotic, lustful, and hungry for more power. She attempted to gain control of the underworld, which was ruled by her sister Ereshkigal. Unfortunately for Inanna, she failed and was killed by her sister's disciple. Without Inanna, the earth withered. But with the assistance of Ea, she and her consort Dumuzi returned to the realms of heaven and earth to restore fertility and rebirth to the natural world. Dumuzi had to spend six months of every year in the underworld as payment for Inanna's resurrection. This cycle corresponded with the annual dormant seasons of autumn and winter on Earth.

Interestingly, the epic poem *Gilgamesh* contains within it a story of a great flood. The Gods decide to eliminate or greatly reduce humanity through a deluge of water. Like the story of Noah's ark, some humans and many animals survive. However, the main theme that emerges from ancient Mesopotamia is that of the fragility of human existence, which is dominated by ruthless and vengeful Gods. There appears to be little confidence in a purposeful afterlife; and even mortal existence is primarily to serve the Gods and take pleasure where you can get it. For the most part, the Gods have little compassion for humanity, with the possible exception of Inanna, who, herself, is prone to violence and disorder. This picture of mortal misery must, however, be contrasted against the advancements in mathematics, astronomy, architecture, art, and science that also characterised this early civilisation.

Greece

The main story of creation in ancient Greek culture comes from the writer Hesiod, who most likely transcribed much earlier oral traditions. As in other creation myths, the universe began with chaos. Out of chaos appeared Gaea (earth), the first creation Goddess, and the formation Gods Tartaros (the depths), Eros (love), Nyx (night), and Erebus (darkness).

Gaea was a feminine representation of the divine. She gave birth to a son, Ouranos, or Uranus. Gaea mated with her son and gave birth to the twelve Titans. The youngest Titan, Kronos, or Cronos, attacked a jealous Ouranos and castrated him. His discarded genitals became the Goddess Aphrodite. Also created from this event were a race of giants and creatures called the Furies. Kronos now became the dominant male God of creation. He mated with his sister Rhea, one of the female Titans, and she gave birth to twelve children. However, a jealous Kronos pre-emptively killed each child by swallowing them at birth. Only the last born, Zeus, escaped this fate. Zeus tricked Kronos into releasing his eleven siblings and together they waged a war of revenge against their father. The remaining Titans, with the exception of Prometheus and Oceanus, rebelled against Zeus. The Titans were finally defeated, and Atlas was condemned to bear the heavens on his shoulders. Zeus and his alliance of Gods now reigned over the divine realm and the earthly world from Mount Olympus as the twelve Gods of Olympus.

The Gods of Olympus and Hades

	God or Goddess	Main Role
1	Zeus	Supreme father of the Gods
2	Hera	The wife of Zeus, Goddess of women and family
3	Athena	Virgin Goddess of wisdom and war
4	Poseidon	God of the sea
5	Apollo	God of the sun and of prophecy
6	Artemis	Goddess of wisdom, war, and Athens
7	Demeter	Goddess of the home
8	Hermes	Messenger of the Gods and healer of the sick
9	Aphrodite	Goddess of beauty and love
10	Ares	God of war
11	Hephaestus	God of fire
12	Hestia	Goddess of the home
Underworld	Hades	God of the underworld (also the name given to the underworld)

It should be noted that various local cultures would include their own preferred Gods, such as Heracles and Dionysus, in the group of elite Gods. The full Greek pantheon was a complex collection of Gods, demigods and God-like humans that had, in most cases, a specified purpose or role to fulfil in the divine or earthly realm. The powers of the Gods were limited in that they could not break a promise and they were not omnipotent. The Gods were also subject to the follies of fate and the actions of others.

There is no one specific story of the creation of humanity in ancient Greece. However, there are two lasting myths, one from Hesiod's works and one from the later Latin author Ovid. These detail two quite different versions of how we came to be. In the first myth, it is the Titan Prometheus who creates man out of mud. This act was done without the permission of Zeus, who commands that man must make sacrifices to the Gods. Zeus also creates Pandora, the first woman, who in turn disobeys the Gods by opening a forbidden box. Humanity is then afflicted with a multitude of diseases and evils as well as the deceptive heart and lying nature of Pandora.

An alternative myth regarding the creation of humanity revolves around different 'ages of creation.' The First, or Golden, Age, occurred during the rule of Kronos. The men who were created at this time were happy and near-perfect. Illness and pain were non-existent. The next age of man was the Silver Age. Man was no longer as content as he had been, and had to work in order to survive. Then Zeus created the men of the Bronze Age, who lived in a warrior-based culture. In this era, war was preferred to civilised peace. Next was the Heroic Age, which included the great adventures of Greek heroes such as Heracles, Theseus, Odysseus, and Achilles. Finally, Zeus created the men of the Iron Age. The men of this era were lazy, depraved, and without honour. Disgusted by this version of man, Zeus decided to destroy humanity by inflicting a great flood to drown all mankind. However, Prometheus's son Deucalion and his wife Pyrrha were warned by their father and escaped the ravages of the flood. Deucalion made sacrifices to Zeus and prayed for the restoration of humanity. Mankind was reborn from the stones thrown by Deucalion (which created men) and those thrown by his wife (which created women).

Rome

Most of what we know about how the early Romans viewed the beginning of the universe comes from the epic poem *Metamorphoses* by Ovid. It was written in the early years of the first millennium. Ovid describes the ages of man as descending from an original creator God who makes a world of order from the seas of chaos. The story of creation is essentially the same as that of the ancient Greeks, with an early golden age that descends to a point where man is destroyed by a catastrophic flood. However, the names of the Greek Gods are changed to those of Roman deities. In particular, Zeus becomes Jupiter, the leader of all the Gods.

	Greek God/Roman God		
1	Zeus	Jupiter	Supreme father of the Gods
2	Hera	Juno	The wife of Zeus/Jupiter, Goddess of women and family
3	Athena Minerva		Virgin Goddess of wisdom and war
4	Poseidon Neptune		God of the sea
5	Apollo	Apollo	God of the sun and of prophecy
6	Artemis	Diana	Goddess of wisdom and war
7	Demeter	Ceres	Goddess of the home
8	Hermes Mercury		Messenger of the Gods and healer of the sick
9	Aphrodite	Venus	Goddess of beauty and love
10	Ares	Mars	God of war
11	Hephaestus	Vulcan	God of fire
12	Hestia	Vesta	Goddess of the home
	Hades	Hades	God of the underworld

An original Roman myth describes the creation of the great city of Rome. A royal priestess known as Alba Longa is made pregnant by the God Mars. She gives birth to twin boys named Romulus and Remus. She is ordered to throw the twins into the River Tiber. However, the babies do not drown; they are carried down the river to a point that would later become the city of Rome. They are rescued by a mother wolf, who rears the boys with her own cubs. Eventually, together the twins create the city of Rome. The relationship between the brothers deteriorates, and Romulus murders Remus to become the sole ruler of Rome. Romulus rules Rome for four decades. Upon his death, he ascends to the divine realm.

Another story that describes the beginning of Rome is that of the *Aeneid*. This poem describes the journey, after the Trojan War, of the defeated warrior Aeneas as he goes in search of a new home. He and other Trojan refugees eventually settle in the area of Rome, already populated by the original descendants of the mythical Romulus. The blending of these two creation stories provides a kind of moral legitimacy for later Roman rule and empire. Romulus is of royal descent, and the warriors of Troy are a direct link to the Hellenic golden age of heroes, who fought for honour, glory, and immortality.

Appendix B: Judeo/Christian and Gnostic Creation Myths

The Creation God of the Hebrew Bible (the Old Testament)

According to the Book of Genesis, the God of the Old Testament created the earth from a void of darkness and nothingness. In six days, God created the earth, the sky, all living things and finally man and woman in his likeness. "So God created Man in his own image, in the image of God, he created him; Male and Female he created them"(Genesis 1:27). A second account, in Genesis 2:4–25, gives a much more detailed description of both the creation and the purpose of the first humans. Adam is made from the dust of the earth and Eve is created from Adam's rib. They live as perfect human beings in the Garden of Eden. Then Eve is tempted by a talking serpent to eat from the forbidden tree of knowledge. She is told that if she does this she and Adam will have knowledge of good and evil and therefore become like God himself. The serpent proclaims to Eve in Genesis 3:5, "For God knows that when you eat of it your eyes will be opened, and you will be like God, knowing good and evil."

Adam, Eve, and all humanity are cursed by God, who declares that, "You shall crawl on your belly, and eat dust all the days of your life" (Gen 3:14). This creation myth does not, however, seem to fit with the idea of an all-knowing and compassionate God. Why would he punish the first humans for simply seeking knowledge about the true nature of things around them? Mysteriously, God says of the fallen Adam, "Adam has now become like one of us, making himself judge of good and evil" (Gen 3:22). This statement could be taken to imply that not only is there more than one God, but that man himself has become like God.

Adam and Eve were forced out of Eden where they give birth to two sons, Cain and Abel. Cain killed Abel and Eve gave birth to another son, Seth. The offspring of Adam and Eve continued for generations but their progress was deemed wicked by God. He decided to eliminate mankind from the earth by means of a great flood. Only Noah, his family, and pairings of animals were spared from the flood, which lasted 40 days and nights. Humanity began again from the offspring of Noah and his family. Civilisation progressed; but humanity was once again punished by God for attempting to build the Tower of Babel, which they thought would make them closer to the divine. Humanity was cursed to speak different languages and to be dispersed across the earth. Eventually, God chose Abraham to settle his people in the land of Canaan. With that, the journey of the Jewish people and their personal relationship with God began.

Judaism has long been acknowledged as a monotheistic religion. This was certainly true by the 6th century BCE. Isaiah 45:5 says, "I am the lord and there is no other" in the Hebrew Bible. However, there are inferences to be made from within the scriptures and from the broader writings of the Talmud that indicate that ancient Judaism was henotheistic, at least in practice. Other gods did exist but were not worshipped by the Jewish people. This is specifically stated in the first Commandment: "You shall have no other Gods before me." This statement suggests that there are other gods but they are specifically to be ignored. The Book of Genesis describes how the "sons of God" take possession of the "daughters of men" (Genesis 6:4) to produce children that are literally like the demigods found in the later Greco-Roman tradition.

The pure monotheistic vision is blurred further by spirit beings, such as angels and other sons of God, who may take human form, as expressed in Job 1:6, where the "sons of God" join God in his council. God himself may appear as an angel messenger or even as a burning bush, as seen by Moses in Exodus 3:2. Further, and like the later Greco-Roman pagans, it is possible for humans to become angels. This was the case for some Jewish martyrs after the first rebellion and the destruction of the Second Temple in around 70 AD. Finally, in Job 1, Satan is seen as an angel or a "son of God," although an adversary. However, God is sovereign over all divine and non-divine beings.

Even God's words, as expressed in the Hebrew Bible, could exist apart from God but also be 'of God.' Like the ancient Egyptians, the Hebrews endowed God-like powers and attributes on their leaders and kings. Moses himself was a representation of God through miracles, and, in Exodus 4.16, even functioned as "if you were God to him" for his brother Aaron. According to the ancient historian Philo, Moses possessed divine attributes as the most perfect human and acted as the "God and King of an entire Nation."

Interestingly, a literal view of the Old Testament can be used for a precise dating of the age of the earth. The lifespans of people in the Bible are incredibly long. Adam lived 930 years; his son Seth, 912 years; the oldest person in the Old Testament, Methuselah, died when he reached the age of 969 (Gen 5:1–27). From these genealogies, along with associated Hebrew and other historical sources, creationist Christians attempt to work out the age of the universe. So, in creationist scripture, the earth is about 6,000 years old.

Genealogy	Timeframe
Before Adam	5 days
From Adam to Noah	1,656 years
From Noah to Abraham	300 years
From Abraham to Jesus	2,000 years
From Jesus to now	2,020 years
Age of the earth	5,976 years

On the other hand, a liberal interpretation of Genesis might conclude that the authors deliberately chose to construct the book in a way that not only provides a simple surface story but also creates an abstract mythology that encourages readers to create alternative or deeper interpretations of the sacred text. Therefore, the words of the Old Testament can be reinterpreted to fit changing conditions in culture and context over time. It has only been in recent times that a literal view of the Bible has been accepted as valid.

The traditional Christian interpretation of the Book of Genesis and the story of creation rests upon the goodness of God and all his creations. The mission of Jesus Christ was to save humanity by providing a new path to faith and redemption, one that rejected the choice of sin. Indeed, the only flaw in humanity was this possibility that one might choose to sin. A core belief of Christianity is that all people inherit the original sin of Adam and Eve and that they will, at some point, choose to sin themselves, bringing evil into the domain of humanity. Only two people, Jesus and his mother Mary, have escaped original sin, as expressed by the tenet of Immaculate Conception.

The meaning of the term 'Immaculate Conception' has been widely misinterpreted. Many people believe it refers to the conception of Jesus by Mary without the need for sexual intercourse. But for the conception to be immaculate, Mary herself needed to be born without original sin in order to be a pure vessel that could hold the future Saviour. The human Jesus was crucified; in the sacrifice of his life, he showed us the path to salvation. Humanity has the choice to abandon evil and surrender itself to the essential goodness of God and his creations in the material world. The Christian Church today still stands firm in the belief that the world in which we live is of value and beauty.

A Gnostic Story of Creation

In 1945, thirteen leather-bound papyrus books dating from around the 2nd century were found by farmhands at Nag Hammadi in Egypt. These books contained 46 works that became known as the Nag Hammadi Library. Most of these works (dated from around the 2nd century CE) were written by Gnostics, whose name comes from the Greek word *gnosis* or 'knowledge.' The Gnostic texts include a variety of myths that attempt to explain the presence of God and the creation of earthly existence. The following discussion will describe both a broad story of creation that concurs with most of the Gnostic scriptures and also a more complex myth of creation from the Nag Hammadi text known as the Secret Book of John.

A consensus approach could conclude the following: originally, there was one unknowable God of pure spirit, the Monad. From this sole divine being emanated other Gods, *æons*, who existed in a spirit realm. One of these divine beings, Sophia, or wisdom, became separated from the spiritual world and gave birth to inferior, possibly evil, divine beings. One of these was the God of the Old Testament, the Hebrew God Jehovah. The Gnostics called this God Ialdabaoth, and they believed he trapped the spirit within the body in the realm of earthly existence. The goal of the Gnostics was to free the enslaved spirit from a false and evil material existence. They believed that because each person had within them a spark of the divine essence, they were superior to Ialdabaoth, who was only a kind of craftsman God and was ignorant of anything he did not create. Therefore, the material world was not a good creation and the craftsman God was himself flawed; he was a demiurge who was unaware of the true divine power that permeated the universe.

The adherents of this story of God and creation believed that the creature who tempted Eve was trying to provide her with knowledge from the tree of *gnosis* about the true nature of the material world and human existence. The mission of Jesus was also to provide true knowledge about the flawed and possibly malevolent nature of the craftsman God and the material world he created. Jesus demonstrated the path to seeking personal *gnosis*, with which an individual could release his or her own divine spark from the entrapment of material existence.

Another early Gnostic story of creation comes from the *Secret Book of John*, from around 150 CE. The first copy of the *Secret Book of John* was found in 1896 but it was not published until 1955. There are only four copies of the book still in existence, of which three are preserved as part of the Nag Hammadi Library. The text contains a revelation by the Saviour to the apostle John, but it is acknowledged that the actual author was not John himself. The *Secret Book of John* describes a supreme divine being so far removed from humanity that it cannot be comprehended by limited human capabilities. At best, it can be understood as an incarnation of pure intellect at peace with itself and alone – the "invisible virgin spirit." This divine entity created the *aeon* forethought, or Barbelo, a feminine manifestation of the invisible spirit. The existence of forethought led to the appearance of other *aeons*: prior acquaintance, incorruptibility, eternal life, and truth. These emanations progressively made the eternal unknowable more tangible. Each *aeon* was composed of a complementary male/female element.

The invisible spirit then "gazed into Barbelo" and created from their union the anointed one, who would be incarnated as Jesus. Below the anointed one were a variety of luminaries, including the word, perception, peace, and wisdom. The female aspect of the divine luminary wisdom is called Sophia. From a thought of Sophia's, the flawed God of the Old Testament, Ialdabaoth, was created. This creation was flawed because it emanated from a lesser divine being without the consent of either her male aspect or the divine spirit itself. Ialdabaoth in turn created lesser spirit beings as well as the material universe that contained humanity.

The story of creation in the *Secret Book of John* acknowledges some of the creation events in Genesis but it differs in its interpretation of what occurred. In this version of the narrative, Ialdabaoth was tricked into creating humanity by a repentant Sophia who was being helped by Barbelo. In this Gnostic understanding of human creation, Adam was created twice. Ialdabaoth created a spiritual Adam in his own image. But because Ialdabaoth himself was a flawed creation, the spiritual Adam was unable to move. Sophia once again tricked Ialdabaoth into "breathing his spirit" into Adam. But it was not the spirit of Ialdabaoth that was given to Adam; rather, it was that of the invisible spirit. Ialdabaoth realised that he had lost his own link with the invisible spirit and that the risen Adam was more powerful than himself. In retaliation, Ialdabaoth and his evil angels trapped the enlightened spiritual Adam in a material body made from the dust of the earth.

This division between body and spirit is central to the Gnostic religious tradition. Humanity has within it an essence of the pure divine that is trapped within an earthly body, and a material universe created by an ignorant, malevolent, and flawed demiurge.

In the *Secret Book of John*, biblical descriptions of the Garden of Eden and the great flood are also reinterpreted through the lens of Gnostic belief. The tree of knowledge was really the tree of *gnosis* and it was from this plant that the true hidden knowledge of the invisible spirit could be gained. In the guise of an eagle, afterthought encouraged Adam and Eve to eat fruit from the tree of *gnosis*. This act enlightened their thinking and "Insight appeared to them as light and awakened their minds." In a fit of jealous rage, Ialdabaoth expelled Adam and Eve from the Garden of Eden. As a sign of his dominance, Ialdabaoth also raped Eve; her sons Cain and Abel were born as imperfect creations. The first purely human creation was Eve's third child, Seth, who was Adam's firstborn son. Some Gnostic sects were known as the sons of Seth. Later, the vengeful Ialdabaoth decided to destroy all humanity for its disobedience by inflicting a flood on the earth. Barbelo warned Noah, his family, and others known as the "immovable race." These people were saved from the wrath of Ialdabaoth.

Interestingly, some forms of early Christianity also rejected the God of the Old Testament. They believed that the hostile and inconsistent God of the Hebrew Bible was incompatible with loving representations of the divine as expressed in the New Testament. This idea is reflected in the teachings of Marcion, an early Christian theologian. His followers not only avoided Judaism; they actually rejected the founding faith of the Old Testament. It was therefore very important that the emerging orthodox church create and endorse a set of scriptures that were united in form, context, and meaning. This set of texts became known as the New Testament, a new agreement or covenant with God.

Endnotes

CHAPTER 1:

1. Otto, S. *Sexual Reproduction and the Evolution of Sex*. Nature Education 1(1):182, <https://www.nature.com/scitable/topicpage/sexual-reproduction-and-the-evolution-of-sex-824/> Accessed June, 2020.

2. Camperio, C. Pellizzari, E. *Fecundity of paternal and maternal non-parental female relatives of homosexual and heterosexual men*, < https://www.ncbi.nlm.nih.gov/pmc/articles/PMC3515521/ > Accessed May, 2020.

3. Willyard, C. *Expanded human gene tally reignites debate After 15 years, researchers still can't agree on how many genes are in the human genome*. Nature research journal, < https://www.nature.com/articles/d41586-018-05462-w > Accessed December 2019.

4. Avise, J. *The Genetic Gods*, Harvard University Press, Massachusetts, USA, 1998, pp. 76-77.

5. Dawkins, R. *Science in the Soul*, Transworld Publishers, London, UK, 2017, p. 145.

6. Avise, J. *The Genetic Gods*, p.148.

7. Starr, D. *Two psychologists followed 1000 New Zealanders for decades. Here's what they found about how childhood shapes later life,* Science Magazine, https://www.sciencemag.org/news/2018/02/two-psychologists-followed-1000-new-zealanders-decades-here-s-what-they-found-about-how> Accessed January, 2020. See also, Polderman, T. Benyamin, B, Leeuw, C, *Meta-analysis of the heritability of human traits based on fifty years of twin studies, < https://www.researchgate.net/publication/276922271_Meta-

analysis_of_the_heritability_of_human_traits_based_on_fifty_years_of_twin_studies > Accessed on March, 2020.

8. Manseau, M. Goff, D. *Cannabinoids and Schizophrenia: Risks and Therapeutic Potential,* Neurotherapeutics, < https://link.springer.com/article/10.1007/s13311-015-0382-6 > Accessed on March 2020.

9. Avise, J. *The Genetic Gods*, pp.200-202.

10. Dawkins, R. *Science in the Soul*, pp. 36-38.

11. Pak, E. *CRISPR; A game-changing genetic engineering technique,* Harvard University Science news, < http://sitn.hms.harvard.edu/flash/2014/crispr-a-game-changing-genetic-engineering-technique/ > Accessed on October 2020.

12. Diacovo, T. Vockley, G. *Sequencing the genome of newborns in the US: Are we ready?* Medical Express, < https://medicalxpress.com/news/2019-07-sequencing-genome-newborns-ready.html > Accessed on June 2020.

13. Zabaneh, D. (lead researcher), *A genome-wide association study for extremely high intelligence,* Molecular Psychiatry, Nature Magazine, < https://www.nature.com/articles/mp2017121 > Accessed September, 2020.

CHAPTER 2:

1. Nicholson, N. *How Hardwired is Human Behaviour?* Harvard Business Review, July/August edition, 1998.

2. Dawkins, R. *Science in the Soul*, Transworld Publishers, London, UK, 2017, p. 145.

3. Avise, J. *The Genetic Gods*, Harvard University Press, Massachusetts, USA, 1998, pp. 154-55.

4. Ibid.

5. Darwin, C. *On the Origin of Species by means of Natural Selection*, Broadview Press. Plymouth, UK. 2003, pp. 537-38.

6. Avise, J. *The Genetic Gods*, pp. 158-59.
7. Dawkins, R. *Science in the Soul*, pp. 58-59.
8. Ibid., p.65.
9. Cosmides, C. Tooby, J. *Evolutionary Psychology: New perspectives on Cognition and Motivation*, Annual Review of Psychology, January. 2013, https://www.annualreviews.org/doi/full/10.1146/annurev.psych.121208.131628 > Accessed on November 2019.

10. Fleishman, D. *Integrating Evolutionary Psychology and Behaviourism*, Dianaverse, April 2020, < https://dianaverse.com/2020/04/26/evpsychandbehaviorism/ > Accessed on August 2020.

11. Fleishman, D. *Universal morality is obscured by evolved morality*, The Evolution Institute special publication, < https://evolution-institute.org/universal-morality-is-obscured-by-evolved-morality/ > Accessed May 2020.
12. Ibid.
13. Curry, O. Mullins, D. Whitehouse, H. *Is It Good to Cooperate? Testing the Theory of Morality-as-Cooperation in 60 Societies*, Current Anthropology Volume 60, Number 1, February 2019.
14. Ibid.
15. Harris, S, *The Moral Landscape*, Transworld Publishers, London, UK, 2010, p. 25.
16. Ibid., p. 44.
17. Ibid., p. 88.
18. Ibid., p. 96.
19. Ibid., p. 123.
20. Ibid., p 144.
21. Dawkins, R. *Science in the Soul*, pp. 39-40.
22. Ibid., p 276.
23. Freud, S. *Civilisation and its Discontents*, W. W. Norton & Company, New York, USA, 2010, p. 145.
24. Ibid., p. 142.

25. Ibid., p. 111.

26. United Nations, *The Universal Declaration of Human Rights (UDHR)*, < https://www.un.org/en/universal-declaration-human-rights/ > Accessed October, 2020.

27. Cosmides, L. Tooby, J. *Universal Minds Explaining the New Science of Evolutionary Psychology,* < https://www.cep.ucsb.edu/155/universalminds.pdf > Accessed on May 2020.

28. McNeil, T. (interview with Daniel Dennett), *Our brain, ourselves,* Tufts Now magazine, < https://now.tufts.edu/articles/daniel-dennett-our-brains-our-selves >

29. Harris, S. *The Moral Landscape*, Transworld Publishers, London, UK, 2010, p. 13.

30. Dennett, D. *The Psychology of Thinking About Free will,* Tufts Education, < https://ase.tufts.edu/cogstud/dennett/papers/psychology_free_will.pdf > Accessed on November, 2020.

31. Ibid.

32. Ibid.

33. Price, M. *Could Morality Have a Transcendent Evolved Purpose?* Psychology Today, < https://www.psychologytoday.com/us/blog/darwin-eternity/201712/could-morality-have-transcendent-evolved-purpose >Accessed on April 2020.

CHAPTER 3:

1. Teilhard de Chardin, P. (Introduction by Julian Huxley), *The Phenomenon of Man*, Wm. Collins & Co, London, UK, 1995, p. 13.
2. Dawkins, R, *Science in the Soul*, Transworld Publishers, London, UK, 2017, p. 124.

3. Darwin, C. *The Origin of Species*, Gramercy Books, New York, USA, 1979. P.459
4. Dawkins, R. *Science in the Soul*, p. 61.
5. Ibid., pp. 297-300.
6. Rothman, J. (interview with Daniel Dennett), *Daniel Dennett's Science of the Soul*, The New Yorker, March 27, 2017, < https://www.newyorker.com/magazine/2017/03/27/daniel-dennetts-science-of-the-soul > Accessed April 2020.
7. Dawkins, R. Wong, Y. *Epilogue To The Mouse's Tale — On "Epigenetics"*, Richard Dawkins Foundation web site, < https://www.richarddawkins.net/2016/06/epilogue-to-the-mouses-tale-on-epigenetics/ > Accessed on December 2019.
8. Dawkins, R. Dennett, D. Harris, S. Hitchens, C. *The Four Horsemen*, Penguin, London, UK, 2019, P. XIX.
9. Ibid., p22.
10. Lents, N. H. *Human Errors*, Weidenfeld and Nicolson, London, UK, 2020. P. 194.
11. Teilhard de Chardin, P. (Introduction by Julian Huxley*)*, *The Phenomenon of Man*, Wm. Collins & Co, London, UK, 1995, p. 11.
12. Campbell, J. Price, M. *Universal Darwinism and the Origins of Order,* Draft Paper, < http://people.brunel.ac.uk/~systmep/Campbell_&_Price_Online_version_1.pdf > Accessed October 2020.
13. Ibid.

14. Price, M. *Entropy and Selection*, Complexity Journal, < https://www.hindawi.com/journals/complexity/2017/47 45379/ > Accessed August 2020.

15. Price, M. *Cosmological Evolution and the Future of Life*, Psychology Today, < https://www.psychologytoday.com/us/blog/darwin-eternity/201802/cosmological-evolution-and-the-future-life > Accessed on September 2020.

16. Ibid.

17. Ibid.

18. Price, M. *Could morality have a Transcendent Evolved Purpose?* Psychology Today, < https://www.psychologytoday.com/au/blog/darwin-eternity/201712/could-morality-have-transcendent-evolved-purpose > Accessed June 2020.

19. Harrison, E. *The Natural Selection of Universes Containing Intelligent Life*, Q. J. R. Astron. Soc., < 1995QJRAS..36..193H Page 193 (harvard.edu) > Accessed June 2020.

CHAPTER 4:

1. Mills, J. Jalil, A. Stanga, P. *Electronic retinal implants and artificial vision: journey and present. Eye,* < https://www.ncbi.nlm.nih.gov/pmc/articles/PMC5639190/ > Accessed on October 2020.

2. Talty, S. *When Artificial Intelligence is everywhere,* Smithsonian Magazine, < https://www.smithsonianmag.com/innovation/artificial-intelligence-future-scenarios-180968403/ > Accessed on April 2020.

3. Dennett, D. *The Age of Artificial Intelligence*, Vernon Press, Delaware, USA, 2020, pp. 27-40.

4. Harari, Y. N. *Homo Deus: A Brief History of Tomorrow,* Vintage Books, London, UK, 2017, p.460.

5. Ibid., pp. 354-55.

6. Ibid., p. 410.
7. Ibid., p. 444.
8. Ibid., pp. 449-50.
9. Tee, J. Taylor, D. *Is information in the brain represented in a continuous or discrete form*, IEEE (online prepublication), < https://drive.google.com/file/d/19PUPICMM0hxeHpTn3OuD eojr_nHvyu4K/view > Accessed on October 2020.
10. Dennett, D. *The Age of Artificial Intelligence*, pp. 59-60.
11. Dennett, D. *Freedom Evolves*, Penguin Books, London, UK, 2003. pp. 6-23.
12. Dennett, D. *The Age of Artificial Intelligence*, p. 30.
13. Ibid., p 49.
14. Reese, B. *The Fourth Age*, Simon & Schuster, New York, USA, 2020, p.36.

CHAPTER 5:
1. Harmer, D. *The God Gene: How Faith is Hardwired into Our Genes*, Random House, New York, USA, 2004, pp. 74-80.

2. Koenig, L. Mcgue, M. Iacono, W. *Rearing Environmental Influences on Religiousness: An Investigation of Adolescent Adoptees,* US National Library of Medicine, < https://www.ncbi.nlm.nih.gov/pmc/articles/PMC2774918/ > Accessed on November 2020.

3. Lewis, G. Bates, T. *The Long reach of the Gene,* The British Psychological Society, < https://thepsychologist.bps.org.uk/volume-26/edition-3/long-reach-gene > Accessed on November, 2020.

4. Kandler, C. *A meta-analytic review of nature and nurture in religiousness across the lifespan,* US National Library of Medicine, < https://pubmed.ncbi.nlm.nih.gov/33068835/ > Accessed on November, 2020.

5. Portnoff, L. McClintock, C. Lau, E. Choi, S. Miller, L. *Spirituality cuts in half the relative risk for depression:*

Findings from the United States, China, and India. Spirituality in Clinical Practice, < https://psycnet.apa.org/record/2017-15879-002 > *Accessed June 2020.*

6. Durkheim, É. *The Elementary Forms of Religious Life*, George Allen & Unwin Ltd., Great Britain, 1915, p. 47.

7. Han, G. Kwang, H. (Review), *Hominin interbreeding and the evolution of human variation*, Journal of Biological research, <https://www.researchgate.net/publication/305364819_Hominin_interbreeding_and_the_evolution_of_human_variation > Accessed September, 2020.

8. Ibid.

9. Dennett, D. *The Age of Artificial Intelligence*, p. 57.

10. Ibid., p, 29.

11. Freud, S. *Moses and Monotheism*, Hogarth Press and the Institute of Psycho-Analysis, 1939, p. 130-131.

12. Ibid., pp. 131–132.

13. Avise, J. *The Genetic Gods*, Harvard University Press, Massachusetts, USA, 1998, p. 206.

14. Cuevas, J. *The Psychological Processes and Consequences of Fundamentalist indoctrination*, Research Gate, < https://www.researchgate.net/publication/276908414_The_Psychological_Processes_and_Consequences_of_Fundamentalist_Indoctrination > Accessed on November 2020.

CHAPTER 6:

1. Robinson, D. *The Great Ideas of Philosophy*, The Teaching Company, 2004, pp. 11-13.

2. Ibid., p, 123.

3. Harkness, G. *The Sources of Western Morality*, Charles Scribner & Sons, New York, USA, 1954, p. 168.

4. Ibid., p. 169.

5. Robinson, D. *The Great Ideas of Philosophy*, pp. 11-14.

6. Harkness, G. *The Sources of Western Morality*, p. 172.

7. Ibid., p. 41.

8. Ibid., p. 181.

9. Ibid., p. 182.
10. Ibid., p. 183.
11. Plato, *The Republic*, Penguin Classics, London, UK, 2007.p.171.
12. Hitchens, C. *God is Not Great*, Allen & Unwin, Sydney, Aust., 2008, p. 309.
13. Armstrong, K. *A History of God*, Vintage Press, London, England, 1999, p. 123.
14. Robinson, D. *The Great Ideas of Philosophy*, p. 126.
15. Armstrong, K. *A History of God*, p. 322.
16. Black, J. *The Secret History of the World*, Quercus, London, UK, 2007, p. 445.

CHAPTER 7:

1. Allegro, J. M. *The Dead Sea Scrolls and the Christian Myth*, Prometheus Books, New York, USA, 1979, p. 232.
2. Tassone, G. *A Study on the Idea of Progress in Nietzsche Heidegger and Critical Theory*, Mellen Press, 2002, USA, p. 323.
3. Huxley, A. *The Doors of Perception*, HarperCollins, New York, USA, 2009, p. 79.
4. Leary, T. *The Psychedelic Experience*, Citadel Press, New York, USA, 1992, p. 13.
5. Ibid., p. 14.

CHAPTER 8:

1. Dawkins, R. *Science in the Soul*, Trans world publishers, London, UK, 2017, p. 34.
2. Ibid.
3. Dennett, D. *If I Ruled the World*, Prospect Magazine, < If I ruled the world: Daniel Dennett | Prospect Magazine > Accessed on November, 2020.
4. Spinoza, B. *Complete Writings of Spinoza: The Ethics, A Theologico-Political Treatise, On the Improvement of*

Understanding, Correspondence, Kindle edition 2018 (locations 8922–8927).

5. Einstein, A. *Ideas and Opinions*, Crown Publishers, NY, USA, 1955, p.38.
6. Ibid., p.37.
7. Ibid., p.40.
8. Ibid., p.40.
9. Spinoza, B. *Complete Writings of Spinoza,* Kindle locations 400–401 and 9025–9027.
10. Robertson, A. *Einstein: A Hundred Years of Relativity*, ABC Books, Sydney, Australia, 2005, p.207.
11. Einstein, A. *Ideas and Opinions*, p.262.
12. Dawkins, R. *The God Delusion*, Transworld Publishers, London, UK, 2016, p.40.
13. Ibid.
14. Dawkins, R. *Science in the Soul*, Transworld Publishers, London, UK, 2017, pp.274-281.
15. Hitchens, C. *God is not Great*, Allen and Unwin, New York, USA, 2007, p.60.
16. Ibid., pp.309–316.
17. Avise, J. The Genetic Gods, Harvard University Press, London, England, 1998, p.208-9.
18. Dawkins, R. *The God Delusion*, p.41.

CHAPTER 9:

1. Harrison, E. *The Natural Selection of Universes Containing Intelligent Life*, Q. J. R. Astron. Soc. p. 194, < 1995QJRAS..36..193H Page 193 (harvard.edu) > accessed December 2020.
2. Ibid., p.195.
3. Ibid.
4. Ibid.
5. Ibid.
6. Reese, B, *The Fourth Age*, Simon & Schuster, New York, USA, 2020, p.31.

7. Harrison, E.R. *Masks of the Universe*, Cambridge University Press, UK, 1985, p. viii.

8. Gould, R. *Universe in Creation*, Harvard University Press, Massachusetts, USA, 2018, p.7.

9. Ibid., p.9.

10. Ibid.

11. Dikian, J. *Quantum Darwinism*, Australian Rationalist, Australian Rationalist Society, September 2020, p.35.

12. Ibid.

13. Hawking, S. *Brief answers to the big Questions*, Bantam Books, New York, USA, 2018, pp. 123-142.

14. Ibid.

15. Ibid.

16. Ibid.

17. Lovelock, J. The Living Earth, Nature Publishing Group, 2003, < 18.12 concepts 769 AM (uv.mx)> accessed January, 2021.

18. Reese, B, *The Fourth Age*, p.240-241.

19. Burton, M. *Ecosystems, from life, to the Earth, to the Galaxy*, School of Physics, University of New South Wales, Sydney, Australia, < 0110694.pdf (arxiv.org) >, accessed on Jan, 2021.

20. Ibid.

21. Bergoglio, Jorge Mario. 'Care for our common home', *The Holy See*, <http://w2.vatican.va/content/francesco/en/events/event.dir.html/content/vaticanevents/en/2015/6/18/laudatosi.html>, May 2015.

22. Vandiver, E. *Classical Mythology*, The Teaching Company, Chantilly, USA, 2000, p.1.

Bibliography

Allegro, J. M. *The Dead Sea Scrolls and the Christian Myth*, Prometheus Books, New York, USA, 1979

Armstrong, K. *A History of God*, Vintage Press, London, England, 1999

Avise, J. *The Genetic Gods*, Harvard University Press, Massachusetts, USA, 1998.

Bacon, F. *On the Advancement of Learning, Books 1 and 2*, Kindle edition, November 2017.

Benyamin, B, Leeuw, C, *Meta-analysis of the heritability of human traits based on fifty years of twin studies* <https://www.researchgate.net/publication/276922271_Meta-analysis_of_the_heritability_of_human_traits_based_on_fifty_years_of_twin_studies >.

Black, J. *The Secret History of the World*, Quercus, London, UK, 2007.

Blainey, G. *A Short History of Christianity*, Penguin Books, Camberwell, Australia, 2011.

Brakke, D. *Gnosticism: From Nag Hammadi to the Gospel of Judas*, Teaching Company, Chantilly, USA, 2015.

Breasted, J.H. *Development of Religion and Thought in Ancient Egypt*, Harper and Brothers Publishers, New York, 1959.

Campbell, J. Price, M. *Universal Darwinism and the Origins of Order*, Draft Paper, < http://people.brunel.ac.uk/~systmep/Campbell_&_Price_Online_version_1.pdf >.

Camperio, C. Pellizzari, E. *Fecundity of paternal and maternal non-parental female relatives of homosexual and heterosexual men*, < https://www.ncbi.nlm.nih.gov/pmc/articles/PMC3515521/ >.

Cosmides, C. Tooby, J. *Evolutionary Psychology: New perspectives on Cognition and Motivation*, Annual review of Psychology, January. 2013,
https://www.annualreviews.org/doi/full/10.1146/annurev.psych.121208.131628 >.

Cosmides, L. Tooby, J. *Universal Minds Explaining the New Science of Evolutionary Psychology,* <
https://www.cep.ucsb.edu/155/universalminds.pdf >.

Cuevas, J. *The Psychological Processes and Consequences of Fundamentalist indoctrination*, Research Gate,
<https://www.researchgate.net/publication/276908414_The_Psychological_Processes_and_Consequences_of_Fundamentalist_Indoctrination >.

Curry, O. Mullins, D. Whitehouse, H. *Is It Good to Cooperate? Testing the Theory of Morality-as-Cooperation in 60 Societies*, Current Anthropology Volume 60, Number 1, February 2019.

Darwin, C. *On the Origin of Species by means of Natural Selection*, Broadview Press. Plymouth, UK. 2003,

Dawkins, R. Dennett, D. Harris, S. Hitchens, C. *The Four Horsemen*, Penguin. Co. London, UK, 2019.

Dawkins, R. Krebs, J.R. 'Arms Races between and within Species', *Proceedings of the Royal Society of London, Series B, Biological Sciences*, Volume 205, Issue 1161, 1979.

Dawkins, R. *Science in the Soul*, Transworld Publishers, London, UK, 2017.

Dawkins, R. *The God Delusion*, Transworld Publishers, London, UK, 2016.

Dawkins, R. *The Greatest Show on Earth*, Transworld Publishers, London, UK, 2010.

Dawkins, R. *The Selfish Gene*, Oxford University Press, New York, USA, 2006.

Dawkins, R. Wong, Y. *Epilogue To The Mouse's Tale — On "Epigenetics"*, Richard Dawkins Foundation web site, < **https://www.richarddawkins.net/2016/06/epilogue-to-the-mouses-tale-on-epigenetics/** >.

deGrasse Tyson, N. Space Chronicles, W.W. Norton & Company, New York, USA, 2012.

Dennett, D. *The Age of Artificial Intelligence*, Vernon Press, Delaware, USA, 2020.

Dennett, D. *Freedom Evolves*, Penguin Books, London, UK, 2003.

Dennett, D. *If I Ruled the World*, Prospect Magazine, < If I ruled the world: Daniel Dennett | Prospect Magazine >,

Dennett, D. *The Psychology of Thinking About Free will*, Tufts Education, < https://ase.tufts.edu/cogstud/dennett/papers/psychology_free_will.pdf >.

Diacovo, T. Vockley, G. *Sequencing the genome of newborns in the US: Are we ready?* Medical Express, < https://medicalxpress.com/news/2019-07-sequencing-genome-newborns-ready.html >.

Durkheim, É. *The Elementary Forms of Religious Life*, George Allen & Unwin Ltd., Great Britain, 1915.

Einstein, A. *Ideas and Opinions*, Crown Publishers, NY, USA, 1955.

Fleishman, D. *Integrating Evolutionary Psychology and Behaviourism*, Dianaverse, April 2020 < **https://dianaverse.com/2020/04/26/evpsychandbehaviorism/** >.

Fleishman, D. *Universal morality is obscured by evolved morality*, The Evolution Institute special publication, < https://evolution-institute.org/universal-morality-is-obscured-by-evolved-morality/ >.

Freud, S. *An Autobiographical Study*, W.W. Norton & Company, NY, USA, [1927] 1952.
Freud, S. *Civilisation and its Discontents*, W. W. Norton & Company, New York, USA, 2010.

Freud, S. *Moses and Monotheism*, Hogarth Press and the Institute of Psycho-Analysis, 1939.

Geraci, R.M. *Apocalyptic Visions of Heaven in Robotics*, Oxford University Press, New York, USA, 2010.

Gould, R. *Universe in Creation*, Harvard University Press, Massachusetts, USA, 2018.

Harari, Y. N. *Sapiens: A Brief History of Humankind*. Vintage Books, London UK, 2014.

Harari, Y. N. *Homo Deus*: *A Brief History of Tomorrow*, Vintage Books, London UK, 2017.

Harkness, G. *The Sources of Western Morality*, Charles Scribner & Sons, New York, USA, 1954.

Harmer, D. *The God Gene: How Faith is Hardwired into Our Genes*, Random House, New York, USA, 2004.

Han, G. Kwang, H. (Review), *Hominin interbreeding and the evolution of human variation*, Journal of   Biological research, <https://www.researchgate.net/publication/305364819_Hominin_interbreeding_and_the_evolution_of_human_variation >.

Harris, S, *The Moral Landscape*, Transworld Publishers, London, UK, 2010.

Harrison, E.R. *Masks of the Universe*, Cambridge University Press, UK, 1985.

Harrison, E.R. *The Natural Selection of Universes Containing Intelligent Life*, Q. J. R. Astron. Soc. < 1995QJRAS..36..193H Page 193 (harvard.edu) >.

Hawking, S. *Brief answers to the big Questions*, Bantam Books, New York, USA, 2018.

Hawking, S., *The Universe in a Nutshell*, Bantam Press, London, UK, 2001.

Hitchens, C. *God is Not Great*, Allen & Unwin, Sydney, Aust, 2008.

Huxley, A. *The Doors of Perception*, HarperCollins, New York, USA, 2009.

Kandler, C. *A meta-analytic review of nature and nurture in religiousness across the lifespan,* US National Library of Medicine, < https://pubmed.ncbi.nlm.nih.gov/33068835/ >.

Koenig, L. Mcgue, M. Iacono, W. *Rearing Environmental Influences on Religiousness: An Investigation of Adolescent Adoptees,* US National Library of Medicine, < https://www.ncbi.nlm.nih.gov/pmc/articles/PMC2774918/ >.

Leary, T. *The Psychedelic Experience*, Citadel Press, New York, USA, 1992.

Lents, N. H. *Human Errors*, Weidenfeld and Nicolson, London, Great Britain, 2020.

Lewis, G. Bates, T. *The Long reach of the Gene,* The British Psychological Society, < https://thepsychologist.bps.org.uk/volume-26/edition-3/long-reach-gene >.

Manseau, M. Goff, D. *Cannabinoids and Schizophrenia: Risks and Therapeutic Potential,* Neurotherapeutics, < **https://link.springer.com/article/10.1007/s13311-015-0382-6** >.

McNeil, T. (interview with Daniel Dennett), *Our brain, ourselves*, Tufts Now magazine, < https://now.tufts.edu/articles/daniel-dennett-our-brains-our-selves>.

Mills, J. Jalil, A. Stanga, P. *Electronic retinal implants and artificial vision: journey and present. Eye.* < https://www.ncbi.nlm.nih.gov/pmc/articles/PMC5639190/ >.

Nicholson, N. *How Hardwired is Human Behaviour*? Harvard Business Review, July/ August edition, 1998.

Otto, S. *Sexual Reproduction and the Evolution of Sex.* Nature Education 1(1):182. <https://www.nature.com/scitable/topicpage/sexual-reproduction-and-the-evolution-of-sex-824/>.

Pak, E. *CRISPR; A game-changing genetic engineering technique,* Harvard University Science News, < http://sitn.hms.harvard.edu/flash/2014/crispr-a-game-changing-genetic-engineering-technique/ >.

Plato, *The Republic*, Penguin Classics, London, UK, 2007.

Portnoff, L. McClintock, C. Lau, E. Choi, S. Miller, L. *Spirituality cuts in half the relative risk for depression: Findings from the United States, China, and India. Spirituality in Clinical Practice,* < https://psycnet.apa.org/record/2017-15879-002 >.

Price, M. *Cosmological Evolution and the Future of Life,* Psychology Today, < **https://www.psychologytoday.com/us/blog/darwin-eternity/201802/cosmological-evolution-and-the-future-life** >.

Price, M. *Could Morality Have a Transcendent Evolved Purpose?* Psychology Today, <

https://www.psychologytoday.com/us/blog/darwin-eternity/201712/could-morality-have-transcendent-evolved-purpose >.

Price, M. *Entropy and Selection*, Complexity Journal, < https://www.hindawi.com/journals/complexity/2017/4745379/ >.

Reese, B. *The Fourth Age*, Simon & Schuster, New York, USA, 2020.

Robertson, A. *Einstein: A Hundred Years of Relativity*, ABC Books, Sydney, Australia, 2005.

Robinson, D. *The Great Ideas of Philosophy*, The Teaching Company, 2004.

Sagan, C. *The Demon Haunted World*, Ballantine Books, New York, USA, 1997.

Spinoza, B. *Complete Writings of Spinoza: The Ethics, A Theologico-Political Treatise, On the Improvement of Understanding, Correspondence,* Kindle edition 2018.

Spivey, N. *Classical Civilisation*, Head of Zeus Publications, London, UK, 2015.

Starr, D. *Two psychologists followed 1000 New Zealanders for decades. Here's what they found about how childhood shapes later life,* Science Magazine, < https://www.sciencemag.org/news/2018/02/two-psychologists-followed-1000-new-zealanders-decades-here-s-what-they-found-about-how>.

Talty, S. *When Artificial Intelligence is everywhere*, Smithsonian Magazine, < https://www.smithsonianmag.com/innovation/artificial-intelligence-future-scenarios-180968403/ >.

Tassone, G. *A Study on the Idea of Progress in Nietzsche Heidegger and Critical Theory*, Mellen Press, New York, USA 2002.

Tee, J. Taylor, D. *Is information in the brain represented in a continuous or discrete form*, IEEE (online prepublication), < https://drive.google.com/file/d/19PUPICMM0hxeHpTn3OuDeojr_nHv yu4K/view >.

Teilhard de Chardin, P. *The Phenomenon of Man*, Wm. Collins & Co, London, UK, 1995.

United Nations, *The Universal Declaration of Human Rights (UDHR)*, < https://www.un.org/en/universal-declaration-human-rights/ >.

Wegner, D. *The Minds best trick: How do we Experience Conscious Will*, Trends in Cognitive Sciences, Vol. 7 N.2 February 2003.

Willyard, C. *Expanded human gene tally reignites debate After 15 years, researchers still can't agree on how many genes are in the human genome.* Nature research journal, < https://www.nature.com/articles/d41586-018-05462-w >.

Zabaneh, D. (lead researcher), *A genome-wide association study for extremely high intelligence,* Molecular Psychiatry, Nature magazine < **https://www.nature.com/articles/mp2017121** >.

INDEX

F

The Obelisk and the Cross
An Alternative History of God, Myth and Meaning in the Western World
Tony Sunderland

This is one of those books that will make you rethink why you believe what you believe. The author writes with simplicity and allows his humorous voice to come across as soothing and very inviting.
Ruffina Oserio for Readers' Favorite

The writing is perfectly concise, incredibly inclusive yet relentlessly focused, and unerringly directed.
Joel R. Dennstedt for Readers' Favorite

Why does an ancient Egyptian obelisk celebrating the God of the sun stand in the centre of St Peter's Square in Vatican City, the home of the Pope and the heartland of Catholicism?

Taking this mysterious fact as his starting point, Tony Sunderland examines the history of religious belief in an attempt to understand how what has happened in the past continues to exert a ghostly influence in the present. Going right back to the voluptuous mother Goddess figures of our ancestors, the pantheons of the Greeks and Romans, the wisdom of the Hebrew Bible, the birth of Christ, the radical heresies of the Gnostics and the Esoterics, the consolidation of a Catholic orthodoxy, and the Protestant Reformation, Sunderland traces a history of ideas that shines a light on how and why belief systems are constructed and the role they play in providing meaning and order in a dangerous, volatile world. Over this long and fascinating history the figure of the first monotheist, the ancient Egyptian pharaoh Akhenaten, casts his long shadow.

Tony Sunderland is an award-winning author and educational researcher who is acknowledged as an innovator in the writing and presentation of nationally accredited courses ranging from social science to the history of learning. He is particularly interested in the practice and history of what has become known as the Western way of life. He believes that there are many alternative explanations of 'how things came to be' in the Western world and that these have either been ignored or suppressed by the dominant and overpowering narratives of consensus history. Tony has visited many of the main archaeological sites in Egypt, Israel, Jordan, Italy, Turkey, and Greece. His current research interests centre on the investigation and understanding of ancient cultures that existed in the vicinity of the greater Mediterranean region. Tony has been married for 31 years and has two children.